PHYSICS RESEARCH AND TECHNOLOGY

QUARK MATTER

FROM SUBQUARKS TO THE UNIVERSE

PHYSICS RESEARCH AND TECHNOLOGY

Additional books and e-books in this series can be found on
Nova's website under the Series tab.

PHYSICS RESEARCH AND TECHNOLOGY

QUARK MATTER

FROM SUBQUARKS TO THE UNIVERSE

HIDEZUMI TERAZAWA

Library of Congress Cataloging-in-Publication Data

ISBN: 978-1-53614-151-1

Published by Nova Science Publishers, Inc. † New York

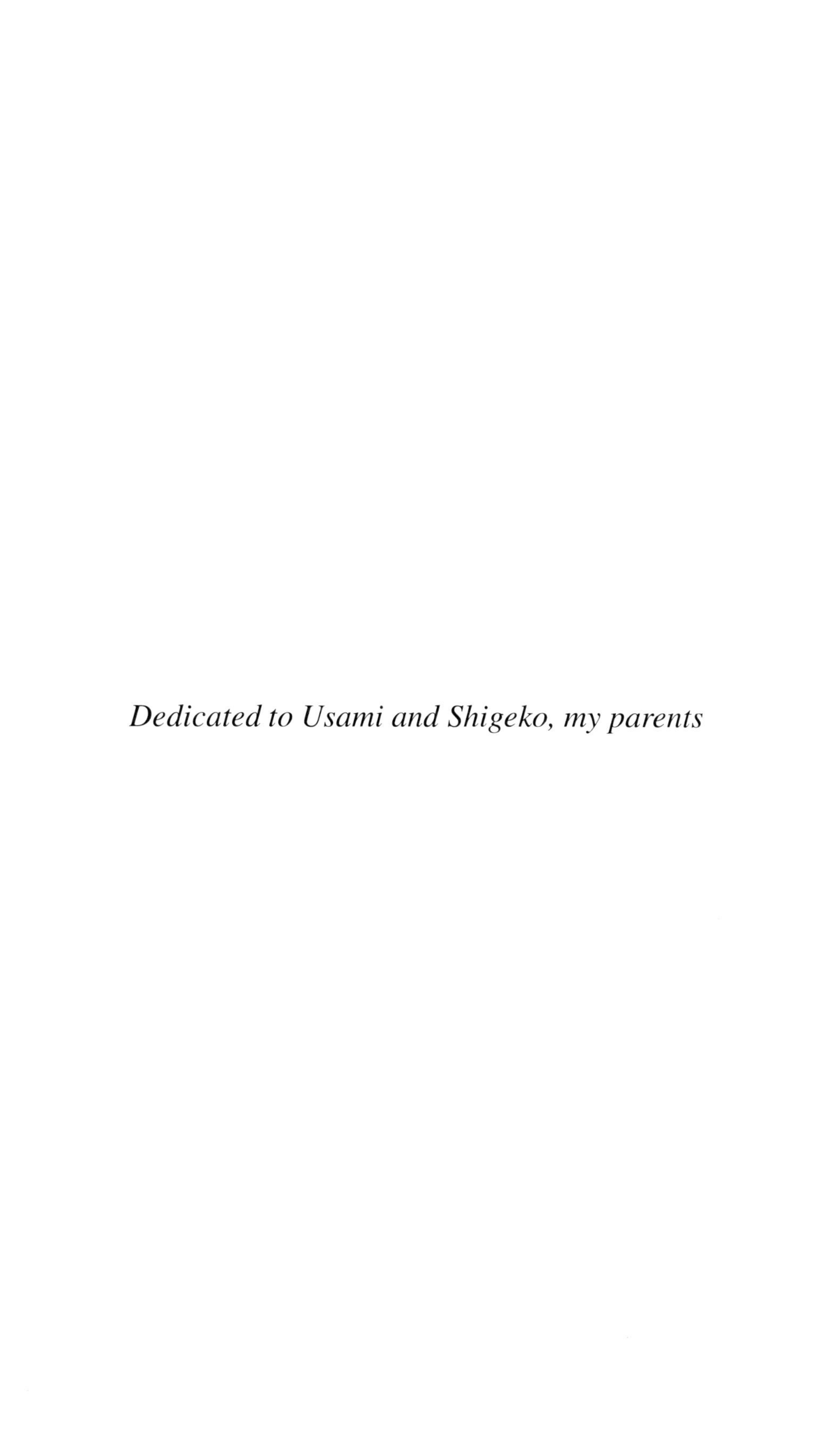

Dedicated to Usami and Shigeko, my parents

Contents

Preface

The meaning of "quark matter" is twofold: 1) compound states of "subquarks" (the most fundamental constituents of matter) of which quarks consist as "nuclear matter" means those of "nucleons" (the constituents of nucleus) of which nuclei consist, and 2) compound states of quarks which consist of roughly equal numbers of up, down, and strange quarks, and which may be absolutely stable. Recently, both of these quark matters have become very intriguing subjects in physics and astronomy since the recently discovered Higgs boson may be taken as a composite object (possibly, a bound state of subquark- antisubquark pairs) and since many recently observed compact stars can be taken as "strange stars" (stars consisting of quark matter). In this book, these hottest subjects in physics and astronomy are discussed without bothering readers with mathematical details.

This book consists of three Chapters: I. Quark Matter and Strange Stars, II. Composites of Subquarks as Quark Matter, and III. Dark Energy, Dark Matter and Strange Stars. Their contents include the followings: In Chapter I, quark matter and strange stars are discussed in detail. In Chapter II, the unified subquark model of all fundamental particles (quarks, leptons, and gauge and Higgs bosons) and forces (strong, electromagnetic, weak, and gravitational forces) is discussed in detail. In Chapter III, pregeometry, in which the general theory of relativity for gravity can be derived as an approximate theory at long distances, is briefly reviewed. Furthermore, special and general theories of "inconstancy" in pregeometry, in which fundamental physical constants may vary, are introduced. And finally, possible solutions to the most puzzling problem in current cosmology of the dark energy and dark matter in the universe are presented. Between Chapter I and Chapter II, a picture of Dr. Abdus Salam and the Author, and interactions with Dr. Abdus Salam and the Author are added as Dr. Salam was one of the founders of subquark models, to be discussed in Chapter II. Also, between Chapter II and Chapter III, a picture of Dr. Andrei Sakharov and the Author, and interactions with Dr. Andrei Sakharov and the Author are added as Dr. Sakharov was the founder of pregeometry, to be discussed in Chapter III.

The readers who want to know more details on individual subjects are recommended to read the original scientific papers quoted in the References of Chapters I, II, and III. Also, those who are not familiar with the **words in bold characters** when they first appear in the texts should find their meanings listed in alphabetical order in the Dictionaries for Chapters I, II, and III. Finally, readers should understand that the "natural unit system" of $c = h/2\pi = 1$, where $c(= 299792458\text{ms}^{-1})$ is the speed of light in vacuum and $h(=6.626070040(81)\times 10^{-34}\text{Js})$ is the Planck constant, is adopted in this book unless otherwise specified.

Acknowledgments

The author would like to thank Mrs. Nadya S. Columbus and the other staff at Nova Science Publishers, Inc. for inviting him to publish a book in the field of "quark matter," and for encouraging him constantly towards completing the manuscript of this book, and to thank Professor Masaki Yasuè for correcting and editing the original manuscripts on which some parts of this book are based. I also thank my parents, Usami and Shigeko Terazawa, to whom this book is dedicated, for everything they did to him for about six decades from 1942 to 2001. Last, but not least, he wishes to thank the many "tutors in his academic life" including Professors Masatoshi Koshiba, Seitaro Nakamura, Hironari Miyazawa, Akito Arima, Keiji Igi, Kazuhiko Nishijima, Yoshio Yamaguchi, Toichiro Kinoshita, Hans Bethe, Abraham Pais, Yoichiro Nambu, and Abdas Salam, for the useful help and valuable advice.

Chapter I

Quark Matter and Strange Stars*

Abstract

New forms of matter such as super-hypernuclei or (strange) quark matter and super-hypernuclear stars or (strange) quark stars are discussed in some detail, based on the so-called "Bodmer-Terazawa-Witten hypothesis" assuming that they are stable (absolutely) or quasi-stable(decaying only weakly).

This Chapter consists of the following three Sections: I. Introduction, II. Exotic Nuclei and Strange Stars, and a Conclusion.

It also contains References and a Dictionary for Chapter I.

*Some contents of this Chapter have been presented in the contributed paper to the XXIII International Seminar, "Nonlinear Phenomena in Complex Systems", Minsk, 2016, "Strange Quark Matter and Strange Quark Stars (*Dedicated to Keiko, my wife*) ", published in the international journal of Nonlinear Phenomena in Complex Systems **19**(2016)147, which is a revised and further up-dated version of the paper entitled, "Exotic Nuclei and Strange Stars (*Dedicated to Professor Masatoshi Koshiba*)", published in the international journal of Nonlinear Phenomena in Complex Systems **18**(2015)25, which was the extended and up-dated version of the Chapters I and III of the contributed paper entitled, "Exotic Matter and Space-Time (*Dedicated to Professor Hans Bethe*)", published in the Proceedings of the 12th International Conference & School "Foundation & Advances in Nonlinear Science", Minsk, Belarus, 2004, edited by V. I. Kuvshinov and G. G. Krylov (Belarusian State University, Minsk, 2005), p.84, and in the Proceedings of the International Conference on New Trends in High-Energy Physics, Yalta, Crimea (Ukraine), 2005, edited by P. N. Bogolyubov, P. O. Fedosenko, L. L. Jenkovszky, and Yu. A. Karpenko (Bogolyubov Institute for Theoretical Physics, Kiev, 2005), p. 259, arXiv:1304.5655v2[physics.gen-ph] 15 Apr 2015.

Between Chapter I and Chapter II, a picture of Dr. Abdus Salam and the Author, and interactions with Dr. Abdus Salam and the Author are added as Dr. Salam was not only one of the founders of the unified gauge theory of electroweak interactions but also one of the founders of the unified composite model of all fundamental particles (quarks, leptons, gauge and Higgs bosons) and forces(strong, electromagnetic, weak, and gravitational forces) to be discussed in Chapter II.

I. Introduction

In the twentieth century, the atomism became one of the most important principles in physics: matter consists of molecules, a molecule consists of atoms, an atom consists of a nucleus and electrons, a nucleus consists of **nucleons**, a nucleon consists of **quarks**, and, perhaps, a quark consists of **subquarks**, the most fundamental constituents of matter [1]. In the twenty-first century, it has become one of the ultimate goals in physics to find the substructure of fundamental particles such as quarks, **leptons**, **gauge bosons**, and **Higgs bosons**. It is, however, still interesting to find new forms of matter which differ from ordinary molecules, atoms, nuclei, hadrons, quarks, and leptons. Recently, such exotic forms of matter as **carbon nanofoams** [2] and **pentaquarks** [3] have been found experimentally (although some doubts in the latter evidence). In the near future, more exotic forms of matter such as **hexalambdas** [4] and **colorballs** [5] would be found. In this Chapter, I am going to discuss these new forms of matter such as **super-hypernuclei** or **(strange) quark matter** and **super-hypernuclear stars** or **(strange) (quark) stars** in some detail, based on the so-called **"Bodmer-Terazawa-Witten hypothesis"** assuming that they are stable(absolutely) or semi-stable (decaying only weakly).

II. Exotic Nuclei and Strange Stars

A super-hypernucleus is a nucleus which consists of many strange quarks as well as up and down quarks. In 1979, I proposed the **quark-shell model of nuclei** in **quantum chromodynamics**, presented the effective two-body potential between quarks in a nucleus, pointed out violent breakdown of **isospin invariance** and importance of **U-spin invariance** in superheavy nuclei, and predicted possible creation of "super-hypernuclei" in heavy-ion collisions at high energies, based on the natural expectation that not only the **Fermi energy** but also the **Coulomb repulsive energy** is reduced in such nuclei [4]. A similar idea was presented independently and almost simultaneously by Chin and Kerman, who called super-hypernuclei "long-lived hyperstrange multiquark droplets"[6]. Five years later in 1984, the possible creation of such super-hypernuclear matter in bulk (or a much larger scale of mass number and space size) in the early Universe or inside **neutron stars** was discussed in detail in QCD by Witten, who called super-hypernuclear matter "quark nuggets" [7] while the properties of super-hypernuclei were investigated in detail in the **Fermi gas model** by Farhi and Jaffe, who called super-hypernuclei "strange matter"[8]. In a series of papers published in 1989 and 1990, I reported an important part of the results of my investigation on the mass spectrum and other properties of super-hypernuclei in the quark-shell model [4].

Let N_u, N_d, and N_s be the number of u's, that of d's, and that of s's, respectively. Then, a nucleus of (N_u, N_d, N_s) has the atomic number, the mass number, and the strangeness given by $Z = (2N_u - N_d - N_s)/3, A = (N_u + N_d + N_s)/3$, and $S = -N_s$. By noting not only a possible similarity between the effective nuclear potential and the effective quark one but also an additional three **color-degrees of freedom**, I have predicted that the magic numbers in the quark-shell model are three times the famous magic numbers in the **nucleon-shell model** $(Z, A - Z = 2, 8, 20, 28, 50, 82, 126, ...)$, i.e., $N_u, N_d, N_s = 6, 24, 60, 84, 150, 246, 378, ...$ Therefore, the magic nuclei such as ${}^4_2He$ and ${}^{16}_8O$ are doubly magic and super-stable also in the quark-shell model since $N_u = N_d = 6$ for ${}^4_2He$ and $N_u = N_d = 24$ for ${}^{16}_8O$. What is new in the quark-shell model is the expectation that not only certain exotic nuclei with a single magic number such as the "dideltas", $D\delta^{++++}$ with $N_u = 6$ and $D\delta^{--}$ with $N_d = 6$, and the "diomega", $D\omega^{--}$ with $N_s = 6$, but also certain super-hypernuclei with a triple magic number such as the "hexalambda", $H\lambda$ with $N_u = N_d = N_s = 6$, and the "vigintiquattuoralambda", $Vq\lambda$ with $N_u = N_d = N_s = 24$, may appear as quasi-stable nuclei. In fact, in the **MIT bag model** [9], one can easily estimate the mass of $H\lambda$ to be as small as 6.3GeV, which is smaller than $6m_\Lambda(\cong 6.7GeV)$. However, in the quark-shell model, there is no qualitative reason why the "dihyperon" or "H dibaryon", H with $N_u = N_d = N_s = 2$, should be quasi-stable or even stable. I have made many other predictions including a sudden increase of the K/π ratio due to production of super-hypernuclei in heavy-ion collisions at high energies [4].

In 1990, Saito et al. found in cosmic rays two abnormal events with the charge of $Z = 14$ and the mass number of $A \cong 370$ and concluded that they may be explained by the hypothesis of super-hypernuclei [10]. In order to find whether these cosmic ray events are really super-hypernuclei as suggested by the cosmic ray experimentalists, I investigated how the small charge-to-mass-number ratio of Z/A is determined when super-hypernuclei are created. In the paper published in 1991, I have shown that such a small charge of $3 \sim 30$ may be realized as $Z \leq \sqrt{2A/3}$($\cong 15.7$ for $A = 370$) if the super-hypernuclei are created spontaneously from bulk super-hypernuclear matter due to the Coulomb attraction [11]. Therefore, the most likely explanation for the abnormal events seems to be that they are, at least, good candidates for super-hypernuclei as suggested by Saito et al. [10].

However, in the other paper published in 1993, I have suggested the second most likely explanation that they may be "technibaryonic nuclei" or "technibaryon-nucleus atoms" [12]. A technibaryon is a baryon which consists of **techniquarks** in a bound state due to the **technicolor force** [13]. A technibaryonic nucleus is a nucleus which consists of nucleons and a technibaryon. A technibaryon-nucleus atom is an atom which consists of a negatively charged technibaryon and an ordinary nucleus in a bound state due to the Coulomb force. The technibaryon mass can be expected to be about 2TeV either from scaling of the baryon mass with the color and technicolor dimensional parameters Λ_C and Λ_{TC} [14] or from my estimation of the techniquark mass to be about $0.5 \sim 0.8$TeV from the **PCDC (Partially-Conserved-Dilation-Current) anomaly sum rule** for quark and lepton masses [15]. The mass value of about 0.4TeV obtained for the abnormal cosmic ray events is much smaller than the expected values for technibaryonic nuclei or technibaryon-nucleus atoms. However, this value would not be excluded since a large experimental error in determining the masses might be involved. In this respect, note that the abnormal cosmic ray event found in 1975 by Price et al. may be better explained by a technibaryonic nucleus or technibaryon-nucleus atom since their later analysis might indicate the charge of $Z \cong 46$ and the mass number of $A \geq 1000$ [16]. Also note that the abnormal cosmic ray event found in 1993 by Ichimura et al. may be better (but much less better) explained by a technibaryonic nucleus or technibaryon-nucleus atom since they reported the charge of $Z \geq 32 \pm 2$ [17].

More recently, I have proposed the third most likely explanation for the abnormal cosmic ray events that they may be "color-balled nuclei " [18]. A color-ball is a color-singlet bound state of an arbitrary number of **gluons** [5] or of "**chroms**", C_α ($\alpha = 0, 1, 2, 3$), which are the most fundamental constituents of quarks and leptons (called "subquarks" in a generic sense) with the color quantum number and which form quarks and leptons together with a weak-isodoublet of subquarks (called "**wakems**"), w_i ($i = 1, 2$), in the unified composite model of all fundamental particles and forces [1]. A color-balled nucleus is a nucleus which consists of nucleons and a color-ball. The color-ball of $(C_0C_1C_2C_3)$ is not only electromagnetically neutral but also weakly neutral. However, it strongly interacts with any hadrons due to the **van der Waals force** induced by the color-singlet state of $(C_1C_2C_3)$ as baryons, the color-singlet states of three quarks. Its mass may be very large as scaled by the subcolor energy scale Λ_{SC} (of the order of, say, 1TeV) [19] and its size may be very small as scaled by $1/\Lambda_{SC}$ ($\sim$1/1TeV) but it may be absolutely stable. This extremely exotic particle (which we may call "primitive hydrogen") may provide us not only the third most likely explanation for the abnormal cosmic ray events but also another candidate for the **missing mass** or **dark matter** in the Universe.

Already in 1970, Itoh and, independently in 1971, Bodmer pointed out the possibility that super-hypernuclei (which the latter called "collapsed nuclei") may exist on a large scale [20]. Bodmer even suggested then that they may explain the missing mass in the Universe. For the last two decades, "strange stars" consisting of super-hypernuclear matter have been investigated in great detail not only theoretically but also experimentally [20]. If the possible identification of the recently discovered unusual **x-ray burster** GRO J1744-28 as a strange star by Cheng, Dai, Wei, and Lu is right [21], the existence of super-hypernuclear matter or "strange matter" has already been discovered by astrophysicists as a gigantic super-hypernucleus or "strangelet", the super- hyperstar or "strange star" in the Universe, before being discovered by high-energy experimentalists in heavy-ion collisions. I must also mention that not only the recent possible identification of the **x-ray pulsar** Her X-1 as a strange star claimed by Li, Dai, and Wang and of the x-ray burster 4U 1820-30 proposed by Bombaci [22] but also the more recent possible identification of the newly discovered millisecond x-ray pulsar SAX J1808.4-3658 suggested by Li, Bombaci, Dey, and van den Heuvel [23] seems to be just as reasonable as that of GRO J1744-28 by Cheng et al. [21]. Very recently, NASA's Chandra X-ray Observatory has found two stars, RXJ185635-3757 which is too small (about 11.3km) and 3C58 which is too cold (less than 1 million degrees in Celsius), being most likely strange stars [24]. As for more recent searches for strange stars, see the reviews by Weber in Ref. [20].

In September, 1999, just before the **BNL RHIC** was about to open up a high-energy range of order hundred GeV/nucleon for heavy-ion-heavy-ion colliding beams, a rumor shocked the whole world [25]. It said that if a negatively charged stable strangelet were produced by RHIC experiments, it would convert ordinary matter into strange matter, eventually destroying the Earth. However, I argued that there is no danger of such a "disaster" at RHIC since the most stable configuration of strange matter must have positive electric charge thanks to the fact that the up quark is lighter than the down and strange quarks [4]. This liberation from such a horrible fear in the "disaster story" would remind us of the fact that the very existence of ordinary matter depends on the mass difference between the proton and neutron which depends on that between the up and down quarks (which further depends on that between the subquarks, w_1 and w_2).

Conclusion

In concluding this chapter, I would like to emphasize that, although there had not yet appeared a clear indication of exotic nuclei such as super- hypernuclei found in high-energy experiments, there are so many candidates for strange matter and strange stars reported by the cosmic-ray experimentalists and by the astrophysicists. Also, I must add that not only the recent discovery of "triaxially deformed nuclei" [26] but also that of "superheavy hydrogen" [27], ${}^5H$ and ${}^7H$, would be something exotic in low-energy nuclear physics.

References and Dictionary

References

[1] Terazawa H., Chikashige Y., and Akama K., *Phys. Rev. D* **15**, 480(1977); Terazawa H., *ibid.* **22**, 184(1980). For a classical review, see Terazawa H., in *Proc. 22nd International Conf. on High Energy Physics*, Leipzig, 1984, edited by Meyer A. and Wieczorek E. (Akademie der Wissenschaften der DDR, Zeuthen, 1984), Vol.I, p.63. For more recent reviews, see Terazawa H., in *Proc. International Conf. "New Trends in High-Energy Physics"*, Alushta, Crimea, 2003, edited by Bogolyubov P. N., Jenkovszky L. L., and Magas V.K. (Bogolyubov Institute for Theoretical Physics, Kiev, 2003), *Ukrainian J. Phys.* **48**, 1292(2003); in *Proc. XXI-st International Conf. "New Trends in High-Energy Physics"*, Yalta, 2007, edited by Bogolyubov P. N., Jenkovszky L. L., and Magas M. K. (Bogolyubov Institute for Theoretical Physics, Kiev, 2007), p.271. For a latest review, see Terazawa H., in *Proc. XXII International Conf. on New Trends in High Energy Physics*, Alushta (Crimea), 2011, edited by Bogolyubov P. N. and Jenkovszky L. L. (Bogolyubov Institute for Theoretical Physics, Kiev, 2011), p.352, arXiv:1109.3705v5[physics.gen-ph] 10 Feb 2012. For our latest papers on unified subquark models of all fundamental particles and forces, see Terazawa H. and Yasuè M., *J. Mod. Phys.* **5**, 205(2014), arXiv:1401.3562v8[hep-ph] 5 Apr 2014; *Nonlinear Phenomena in Complex Systems* **19**, 1(2016), arXiv:1505.00172v2 [hep-ph] 9 Sep 2015; and Terazawa H., *ibid.* **20**, 327(2017).

[2] For a brief review, see, for example, Giles J., Nature Science Update, 23 March 2004, http://www.nature.com/nsu/040322/040322-5.html. See also Minot E. D. et al., *Phys. Rev. Lett.* **90**, 156401(2003), arXiv:cond-mat/0211152v3[cond-mat.mes-hall] 4 Feb 2003; *Nature* **428**, 536(2003), arXiv:cond-mat/0402425v2[cond-mat.mes-hall] 5 Mar 2004.

[3] For the first experimental evidences, see Nakano T. et al. (LEPS Collaboration), *Phys. Rev. Lett.* **91**, 0120002(2003), and Barmin V. V. et al. (DIANA Collaboration), *Phys. Atom. Nucl.* **66**, 1715(2003); *Yas. fiz.* **66**, 1763(2003).

[4] For the prediction of a super-hypernucleus, H_λ (hexalambda), with $B = 6$ and $S = -6$ at $m \cong 5.6 \sim 6.3$GeV in the MIT bag model and at $m \cong 7.0$GeV in the quark-shell model, see Terazawa H., INS-Report-336 (INS, Univ. of Tokyo) May, 1979; *J. Phys. Soc. Jpn.* **58**, 3555(1989); **58**, 4388(1989); **59**, 1199(1990). For a classic review on super-hypernuclei, see Terazawa H., in *Proc. 2nd Conf. on Nuclear and Particle Physics*, Cairo, 1999, edited by Comsan N. M. H. and Hanna K. M. (Nuclear Research Center, Atomic Energy Authority, Cairo, 2000), p.28. For a more recent review, see Terazawa H., in *Proc. International Conf. on New Trends in High-Energy Physics*, Yalta, Crimea (Ukraine), 2005, edited by Bogolyubov P. N., Fedosenko P. O., Jenkovszky L. L., and Karpenko Yu. A. (Bogolyubov Institute for Theoretical Physics, Kiev, 2005), p.259, arXiv:1304.5655v2 [physics.gen-ph] 15 Apr 2015.

[5] For reviews, see Terazawa H., in *Proc. International Conf. (VIIIth 'Blois Workshop') on Elastic and Diffractive Scattering*, Protvino, 1999, edited by Petrov V. A. and Prokdin A. (World Scientific, Singapore, 2000), p.153; in *Proc. IX Annual Seminar "Nonlinear Phenomena in Complex Systems"*, Minsk, 2000, edited by Kuvshinov V. I. et al. (Institute of Physics, National Academy of Sciences of Belarus, Minsk, 2000), *Nonlinear Phenomena Complex Systems* **3**:3, 293(2000). Recently, Hooper et al. have pointed out that the observed 511keV emission from the galactic bulge by INTEGRAL (International Gamma-Ray Astrophysics Laboratory), and previously by Compton Gamma-Ray Observatory, could be due to very light (1-100MeV) annihilating dark matter particles. It can be taken as the first observation of very light color-balls. See D.Hooper et al., *Phys. Rev. Lett.* **93**, 161302(2004). For related works, see Boehm C. et al., *Phys. Rev. Lett.* **92**, 101301 (2004); Picciotto C. and Pospelov M., *Phys. Lett.* **B605**, 15(2005); Hooper D. and Wang L. T., *Phys. Rev. D* **70**, 063506(2004); Casse M. et al., astro-car/0404490. See also Oaknin D. H. and Zhitnitsky A. R., *Phys. Rev. Lett.* **94**, 101301(2005), in which they discuss the possibility that it can be naturally explained by the supermassive very dense droplets (strangelets) of dark matter.

[6] Chin S. A. and Kerman A. K., *Phys. Rev. Lett.* **43**, 1292(1979).

[7] Witten E., *Phys. Rev. D* **30**, 272(1984).

[8] Farhi E. and Jaffe R. L., *Phys. Rev. D* **30**, 2379(1984); **32**, 2452 (1985).

[9] Chodos A., Jaffe R. L., Johnson K., Thorn C. B., and Weisskopf V. F., *Phys. Rev. D* **9**, 3471(1974); Chodos A., Jaffe R. L., Johnson K., and Thorn C. B., *ibid.***10**, 2599(1974); DeGrand T., Jaffe R. L., Johnson K., and Kiskis J., *ibid.***12**, 2060(1975); Jaffe R. L., *ibid.* **15**, 267, 281(1977). For the prediction of a stable "di-lambda ", H, in the MIT bag model, see Jaffe R. L., *Phys. Rev. Lett.* **38**, 195, 612(E)(1977).

[10] Saito T., Hatano Y., Fukuda Y., and Oda H., *Phys. Rev. Lett.* **65**, 2094 (1990); Kasuya M., Saito T., and Yasuè M., *Phys. Rev. D* **47**, 2153(1993). For reviews, see Mori K. and Saito T., in *Proc. 24th International Cosmic Ray Conference*, Rome, 1995 (Rome, 1995), Vol.11, p.878; Saito T., *ibid.*, Vol.11, p.898.

[11] Terazawa H., *J. Phys. Soc. Jpn.* **60**, 1848(1991).

[12] Terazawa H., *J. Phys. Soc. Jpn.* **62**, 1415(1993).

[13] Weinberg S., *Phys. Rev. D* **19**, 1277(1979); Susskind L., *ibid.* **20**, 2619(1979). For the earlier related proposal, see Bég M. A. B. and Sirlin A., *Ann. Rev. Nucl. Sci.* **24**, 379(1974).

[14] See, for example, Farhi E. and Susskind L., *Phys. Rep.* **74**, 277(1981).

[15] Terazawa H., *Phys. Rev. Lett.* **65**, 823(1990). For the earlier works on the PCDC sum rule later leading to the QCD sum rule in a generic sense, see also Terazawa H., *Phys. Rev. Lett.* **32**, 694(1974); *Phys. Rev. D* **9**, 1335(1974); **11**, 49(1975); **12**, 1506(1975).

[16] Price P. B. et al., *Phys. Rev. Lett.* **35**, 487(1975); *Phys. Rev. D* **18**, 1382(1978).

[17] Ichimura M. et al., *Nuovo Cimento* **A106**, 843(1993).

[18] Terazawa H., *J. Phys. Soc. Jpn.* **69**, 2825(2000); in *Proc. IX Annual Seminar "Nonlinear Phenomena in Complex Systems"*, Minsk, 2000, edited

by Babichev L. and Kuvshinov V. (Institute of Physics, National Academy of Sciences of Belarus, Minsk, 2000), *Nonlin. Phenom. Complex Syst.* **9**, 280(2000).

[19] Hooft G. ’t, in *Recent Developments in Gauge Theories*, edited by Hooft G. ’t (Plenum, New York, 1980), p.135; Terazawa H., *Prog. Theor. Phys.* **64**, 1763(1980).

[20] Itoh N., *Prog. Theor. Phys.* **44**, 291(1970); Bodmer A. R., *Phys. Rev. D* **4**, 1601(1971). For a classical review, see Weber F., Schaab Ch., Weigel M. K., and Glendenning N. K., Report No. LBL-37264, UC-413 (LBL, Berkeley, 1995), in *Proc. Ringer Workshop*, Tegernsee, Germany, 1995, presented at Conference: C95-03-06. See also Glendenning N. K., *Compact Stars, Nuclear Physics, Particle Physics, and General Relativity*, 2nd ed.(Springer-Verlag, New York, 2000). For a recent review, see Weber F., arXiv:astro-ph/0407155v2 27 Sep 2004, published in *Prog. Part. Nucl. Phys.* **56**, 193(2005). For a more recent review, see Weber F. et al., arXiv: 1210.1910[astro-ph.SR]6 Oct 2012, published in the Proceedings of the IAU Symposium **291**, edited by van Leeuwen J. (International Astronomical Union, 2013), p.61. For some latest papers on strange matter and strange stars, see Xu J. F. et al., *Phys. Rev. D* **92**, 025025(2015), arXiv:1512.08229v1[hep-ph] 27 Dec 2015, and many references therein.

[21] Chen K. S., Dai Z. G., Wei D. M., and Lu T., *Science* **280**, 407(1998). See also Chen K. S. and Dai Z. D., *Phys. Rev. Lett.* **77**, 1210(1996).

[22] Li X.-D., Dai Z.-G., and Wang Z.-R., *Astron. Astrophys.* **303**, L1(1995); Bombati I., *Phys. Rev. C* **55**, 1587(1997). See also Dey M., Bombaci I., Dey J., Rey S., and Samanta B. C., *Phys. Lett.* **B438**, 123(1998).

[23] Li X.-D., Bombaci I., Dey M., Dey J., and van den Heuvel E. P. J., *Phys. Rev. Lett.* **83**, 3776(1999).

[24] For a review, see http://www1.msfc.nasa.gov/NEWSROOM /news/releases/2002/02-082.html, April 10, 2002 and Seife C., *Science* **296**, 238(2002). For RXJ185635-3754, see Pons J. A. et al., *Astrophys. J.* **564**, 981(2002); Drake J. J. et al., *ibid.* **572**, 996 (2002); Walter F. M. and Lattimer J., *ibid.***576**, L145(2002). For 3C58, see Slane P. et al., *Astrophys. J.* **571**, L45(2002); Yakovlev D. G. et al., *Astron. Astrophys.* **389**, L24(2002).

[25] Marburger J., http://www.pubaf.bnl.gov/pr/bnlpr091799.html, September 17, 1999; http://www.bnl.gov/bnlweb/rhicreport.html, October 6, 1999.

[26] Odegard S. W. et al., *Phys. Rev. Lett.* **86**, 5866(2001).

[27] Korsheninnikov A. A. et al., *Phys. Rev. Lett.* **87**, 092501 (2001); **90**, 082501(2003). For the recent observation of excited states in ^5H, see Golovkov M. S. et al., *Phys. Rev. Lett.* **93**, 262501(2004).

Dictionary

BNL RHIC: The **R**elativistic **H**eavy **I**on **C**ollider which is the accelerator (at **B**rookhaven **N**ational **L**aboratory) for high energy experiments by colliding heavy ions such as *Au* and *U* ions against each other in order to study nuclear physics.

"Bodmer-Terazawa-Witten hypothesis": The hypothesis that (strange) quark matter which is a color-singlet state consisting of almost equal numbers of u, d, and s quarks is (absolutely) stable or at least quasi-stable (decaying only weakly). It is based on the expectation that the energy of such strange quark matter with a certain baryon number, $N_B = (N_u + N_d + N_s)/3$, may be lower than that of an ordinary matter with the same baryon number although the s quark mass is much larger than the u and d quark masses, since not only the Fermi energy but also the Coulomb repulsive energy is much reduced in such state [4, 6, 7, 8, 20]. This hypothesis was proposed by the author (who called strange quark matter "super-hypernuclei") in the quark- shell model of nuclei in QCD (quantum chromodynamics) in 1979 [4].

carbon nanofoams: The large molecules consisting of many Carbon atoms whose shapes have many foam-like structures. In 2003, such exotic form of matter was created and taken as extremely useful material as carbon nanotubes, which are large molecules consisting of many Carbon atoms whose shapes have tube-like structures [2].

chroms: A Pati-Salam-color quartet of subquarks, C_α for $\alpha = 0, 1, 2, 3$, which have the electric charge of $e/2$ and $-e/6$ for C_0 and C_i for $i = 1, 2, 3$, respectively, and which are scalar and bosons. They were first proposed by Pati and Salam in 1974 in their unified model of quarks and leptons, In the unified composite model of all fundamental particles (quarks, leptons, gauge and Higgs bosons, and even gravitons) and forces(strong, electromagnetic, weak, and even gravitational interactions) [1], the fundamental fermions, quarks and leptons, are composites of a wakem (w_i where i = 1 or 2) and a chrom (C_α where $\alpha =$ 0,1,2, or 3) while the gauge and Higgs bosons, and even the gravitons are composites of w and anti-w or C and anti-C pairs. The detailed descriptions of their properties are given in Chapter II.

color-balls: The composite particles which are color-singlet and electrically neutral bound states consisting of an arbitrary number of gluons or of "chroms", C_α ($\alpha = 0,1,2,3$), which are the most fundamental constituents of quarks and leptons (called "subquarks" in a generic sense) with the color quantum number and which form quarks and leptons together with a weak-isodoublet of subquarks (called "wakems"), w_i ($i = 1,2$), in the unified composite model of all fundamental particles and forces [1]. The color-ball of $(C_0C_1C_2C_3)$ is not only electromagnetically neutral but also weakly neutral. However, it strongly interacts with any hadrons due to the van der Waals forces induced by the color-singlet state of $(C_1C_2C_3)$ as baryons, the color-singlet states of three quarks.

color-degrees of freedom: The three degrees of freedom (r,g,b) which quarks have in addition to at least six flavor-degrees of freedom (u,d,s.c,b,t). In the unified model of quarks and leptons of the Pati-Salam type, leptons have the fourth color (l).

Coulomb repulsive energy: The positive potential energy due to the repulsive Coulomb force between two particles with the same sign of electric charges.

dark matter: The matter whose contents have not yet been seen but whose density of the Universe accounts (25.8$\pm$ 1.1)percents of the critical density of the Universe. It is also called "missing mass in the Universe".

Fermi energy: The positive energy of a compound state obtaining due to the presence of constituent fermions, the same kind of which can not be in the same low-energy state in the Fermi statistics.

Fermi gas model: The theoretical model in which some matter is taken as a state of gas consisting of many fermions interacting with each other very weakly.

gauge bosons: The fundamental particles which mediate forces between fundamental particles such as quarks, leptons, and gauge bosons themselves. There are at least a dozen of them: an octet of the gluons (g's) of $SU(3)_c$ for strong interactions between quarks and gluons themselves, the photon (γ) of electromagnetic interactions, and the charged and neutral weak bosons ($W^\pm$ and Z) of weak interactions. All of them are vector particles(having proper spin of

$h/2\pi$) and bosons in statistics The charged weak bosons, $W^{\pm}$, have the electric charge of $q_W = \pm$e while all the other gauge bosons are electrically neutral. The photon and gluons are massless and the weak boson masses are measured to be $m_W = (80.385 \pm 0.015)$GeV/c^2 and $m_Z = (91.1879 \pm 0.0021)$GeV/$c^2$.

gluons: A color-octet of the fundamental particles (g's) mediating strong interactions between quarks, which are described as $A^a_\mu(\mu = 0,1,2,3$ and $a = 1,2,3,...,8)$ in QCD. They are massless, vector(having proper spin of $h/2\pi$), and bosons in statistics.

hexalambdas: The compound nuclei which consist of six Λ baryons or, equivalently, which are color-singlet states consisting of eighteen quarks of 6u's, 6d's, and 6s's. In the unified quark- shell model of nuclei [4], it has been predicted that they may be stable since they are electrically neutral and have magic numbers of 6 triply for all quark numbers(N_q's).

Higgs bosons: The fundamental particles which are scalar (having no proper spins) and generate non-vanishing masses to the weak bosons(W and Z) and fundamental fermions(quarks and leptons) due to the spontaneous breakdown of gauge symmetry of electroweak interactions. In the unified gauge model of Glashow-Salam-Weinberg for electromagnetic and weak interactions of quarks and leptons, a weak-isodoublet of the Higgs bosons, ϕ, are assumed and the neutral Higgs boson, $H(=\phi^0)$, has been found by the LHC (Large-Hadron-Collider) experiments at CERN with the mass of $m_H = (125.04 \pm 0.24)$GeV/c^2, as predicted.

isospin invariance: The rule that the total isospin of an initial state is conserved in any reactions of strong interactions. It is violated both in electromagnetic and in weak interactions.

leptons: The fundamental particles which have no strong interactions. There are at least six flavors of them: three charged leptons including the electron(e), muon(μ), and tau-particle(τ), which have a negative electric charge ($q_e = q_\mu = q_\tau = -$e), and their neutrinos including the electron neutrino(ν_e), mu-neutrino (ν_μ), and tau-neutrino(ν_τ), which are electrically neutral. The pairs of (ν_e,e), (ν_μ,μ), and (ν_τ,τ) form doublets of the weak-isospin of $SU(2)_w$ and are called the first, second, and third generation of leptons. All of them have proper spin of $h/4\pi$ and fermions. The charged lepton masses are measured to be m_e = 0.5109989461(31)MeV/c^2 = 548.579909070(16)$\times$ 10^{-6}u, m_μ = 105.6583745(24)MeV/c^2 = 0.1134289257 (25)u, and m_τ = 1776.86(12)MeV/c^2 while the neutrino masses may not be vanishing but be all extremely small (less than 2eV/c^2), and none of them has not yet been measured.

missing mass: See **dark energy**.

MIT bag model: The theoretical model in which a hadron is taken as a bag containing quarks with their effective masses and with the fixed bag constant for the average energy density of the "confining material".

neutron stars: The stars which consist of neutrons.

nucleons: The fundamental particles of which atomic nuclei consist. There are two kinds of them: the proton (p) which has a positive electric charge ($q_p = e = 1.6021766208(98) \times 10^{-19}C = 4.803204673(30) \times 10^{-10}esu$) and the neutron (n) which is electrically neutral ($q_n = (-0.2 \pm 0.8) \times 10^{-21}e$). Both of them have proper spin of $h/4\pi$, where h is the Planck constant ($h = 6.626070040(81) \times 10^{-34}Js$), and fermions(particles which obey the Fermi statistics in which the one energy state can be occupied by only one particle). Their masses are m_p = $038.2720813(58)MeV/c^2$ = $1.007276466879(91)u$ and m_n = $939.565413(6)MeV/c^2$ = $1.0086649159(5)u$.

nucleon-shell model: The theoretical model of nuclei in which nucleons are taken as particles freely moving in the effective potential of nuclei so that the motion of nucleons in a nucleus has a shell structure similar to that of electrons in an atom.

PCDC (Partially-Conserved-Dilation-Current) anomaly sum rule: The sum rule for quark and lepton masses derived by the author from the Ward identity for the vertex function of the trace of the energy-momentum tensor and the two axial vector currents, and from the partially conserved axial-vector current hypothesis [15].

pentaquarks: The compound states which are color-singlet states, consisting of four up or down quarks and a single up or down anti-quark and having the baryon-number, $N_B = N_q/3 = 1$, as ordinary baryons which are color-singlet states, consisting of three up or down quarks. In 2003, the LEPS and DIANA Collaborations reported the first evidence for a quasi-stable pentaquark state [4] although there were some doubts on their claims at that moment.

quantum chromodynamics (QCD): The Yang-Mills gauge theory of color $SU(3)_c$ which describes strong interactions of quarks.

quarks: The fundamental particles of which hadrons (particles with strong interactions) consist. There are at least six flavors of them: the up (u), charm (c), and top (t) quarks which have a positive electric charge ($q_u = q_c = q_t = 2e/3$), and the down (d), strange (s), and bottom (b) quarks which have a negative electric charge ($q_d = q_s = q_b = -e/3$). The pairs of (u,d), (c,d), and (t,b) form doublets of the "weak isospin"of $SU(2)_w$ and called the first, second, and third generation of quarks. All of them have proper spin of $h/4\pi$, and fermions. Furthermore, all of them have three degrees of freedom called "colors" and are members of a fundamental color-triplet of $SU(3)_c$. Therefore, there must be at least three generations of iso-doublets of three color-triplets of quarks, eighteen quarks all together. There masses are estimated to be $m_u = 2.2 \pm 0.6 GMeV/c^2$, $m_d = 4.7 \pm 0.5 MeV/c^2$, $m_s = 96 \pm 8 MeV/c^2$, $m_c = 1.27 \pm 0.03 GeV/c^2$, $m_b = 4.18 \pm 0.04 GeV/c^2$, and $m_t = 173.21 \pm 0.51 \pm 0.71 GeV/c^2$.

quark matter: See **super-hypernuclei**.

quark stars: See **super-hypernuclear stars**.

quark-shell model of nuclei: The theoretical model of nuclei in which quarks are taken as particles freely moving in the effective potential of nuclear matter as nucleons in the nucleon shell-model of nuclei. It was proposed by the author in 1979, based on the similarity of the effective potential between quarks and that between nucleons except for the additional three color-degrees of freedom possessed by quarks [4].

strangeness: The additive quantum number(S) for hadrons (particles with strong interactions). $S = -N_s$ where N_s is the total number of strange quarks(s's) inside of a hadron. For example, $S = \pm 1$ for the $K^{\pm}$ mesons and $S = -1$ for the Λ and Σ baryons. It is conserved in strong interactions as the number of s quarks is conserved.

strange quark matter: See **super-hypernuclei**.

strange quark stars: See **super-hypernuclear stars**

strange stars: See **super-hypernuclear stars**

subquarks: The most fundamental particles of which quarks (and leptons) consist. In our unified composite model of all fundamental particles(quarks, leptons, gauge bosons, and Higgs scalars) and forces (strong, weak, and electromagnetic interactions) [1], not only the quarks and leptons but also the gauge bosons(the photon, the weak bosons, and the gluons), the Higgs boson, and even the graviton are all composites of subquarks. There are two types of subquarks: a weak-isospin doublet of subquarks (w_i) $(i = 1,2)$ which have electric charge $\pm e/2$ and a color-quartet of subquarks $(C_\alpha)(\alpha = 0,1,2,3)$ which have electric charge $e/2$ or $-e/6$ for $\alpha = 0$ or $\alpha = 1,2,3$, respectively. The former subquarks have proper spin of $h/4\pi$ and are fermions called "wakems"(standing for **we**ak **a**nd electro**m**agnetic) while the latter have no proper spin and are bosons called "chroms" (standing for colors). Furthermore, all of them may have four degrees of freedom called "subcolors" and are members of a fundamental subcolor-quartet of $SU(4)_{sc}$. Therefore, there may be at most twenty-four subquarks including eight weak-isodoublet and subcolor-quartet of wakems and sixteen color-quartet and subcolor-quartet of chroms.

super-hypernuclear stars: The stars which consist of (strange) quark matter. The possibility of such strange stars in the universe was first pointed out by Itoh in 1970 and, independently in 1971, by Bodmer who even suggested that it may explain the missing mass(or dark matter) in the universe [20]. For the last two decades, many astrophysicists including Cheng et al. [21], Bombaci et al. [22], and Li et al. [23] have proposed to identify the newly discovered unusual X-ray bursters, the X-ray pulsers, the too small star and the too cold star as strange quark stars [24]. It should be noticed that one may call the super-hypernuclear stars or strange quark stars "strange stars", but that some authors in literatures called these stars or the other stars consisting of ordinary (up and down) quarks such as neutron stars in the quark-gluon plasma state "quark stars", which might be misleading.

super-hypernuclei: The nuclei which are color- singlet states consisting of almost equal numbers of up, down, and strange quarks. In the unified quark-shell model in QCD (quantum chromodynamics) [4], it has been predicted by the author since 1979 that some of them may be (absolutely) stable or at least quasi-stable(decaying only weakly), which is the so-called "Bodmer-Terazawa-Witten hypothesis" [4, 7, 20]. In 1990, Saito et al. claimed to have found two good candidates for strange quark matter in cosmic rays [10]. More recently, many astrophysicists have claimed good candidates for strange quark stars which consist of strange quark matter [21, 22, 23, 24]. It should be noticed that one may call the super-hypernuclei or strange quark matter simply as "quark matter" but that some authors in literatures called matter consisting of ordinary (up and down) quarks in the quark-gluon plasma state "quark matter", which might be misleading.

technicolor force: The hypothetical force mediated by "techni- gluons" which are the hypothetical gauge bosons in the techniquark model for spontaneous breakdown of $SU(2)_W$ symmetry.

techniquarks: The hypothetical quarks with the hypothetical "technicolor"-degrees of freedom ($Q = U$ and D) which may make spontaneous breakdown of $SU(2)_W$ symmetry by condensation of Q-anti- Q pairs in the vacuum. generating the Higgs bosons as composites of Q- anti-Q pairs.

U-spin invariance: The rule that the total U-spin of a initial state is conserved in a reaction of strong interactions, where the U-spin is the SU(2)-spin rotating in the SU(3)-space of d and s as the isospin is the SU(2)-spin in that of

u and d. It holds even in electromagnetic interactions but breaks down in weak interactions.

van der Waals force: The effective interaction between atoms, molecules, or any compound objects, induced by the fundamental interaction between their constituents.

wakems: A weak-isospin doublet of subquarks, w_i for $i = 1, 2$, which have the electric charge of $\pm e/2$ and proper spin of $h/4\pi$ and are fermions. They were first introduced by the author and his collaborators in 1977 in the unified composite model of all fundamental particles(quarks, leptons, gauge and Higgs bosons, and even gravitons) and forces(strong, electromagnetic, weak, and even gravitational interactions) [1]. The fundamental fermions, quarks and leptons, are composites of a wakem(w_i where $i = 1$ or 2) and a chrom(C_α where $\alpha = 0,1,2$, or 3) while the gauge and Higgs bosons, and even the graviton are composites of w and anti-w pairs or C and anti-C pairs. The detailed description of their properties are given in Chapter II.

x-ray burster: The astronomical object which bursts a gigantic number of X-rays whose total amount of energy is so tremendous that they may be observed by X-ray detectors on the satellites or on the earth.

x-ray pulsar: The astronomical object which emits X-rays in high frequency whose total amount of energy is so tremendous that they may be observed by X-ray detectors on the satellites or on the earth.

Picture of Dr. Abdus Salam (1926-1996) and the Author in the 1st Conference Room, Institute of Nuclear Study, University of Tokyo in 1980.

Interactions with Dr. Abdus Salam and the Author

1) In 1974, A. Salam and J. C. Pati suggested a composite model in which quarks (u_i, d_i, c_i, s_i where $i = 1.2.3$) and leptons (ν_e, e, ν_μ, μ) are made of two "preons", f_a and C_α (where $a = u, d, c, s$ and $\alpha = 1, 2, 3$, and l). A few years later, in 1977, we proposed the unified composite model of all fundamental particles and forces in which not only quarks and leptons but also gauge and Higgs bosons are made of two types of "subquarks", w_i (called "wakems" standing for **we**ak and **e**lectro**m**agnetic) and C_α (called "chroms" standing for colors) (where $i = 1, 2$ and $\alpha = 0, 1, 2, 3$). Later, many others proposed different composite models of quarks and leptons with different names for the more fundamental constituents of matter such as "rishons"by H. Harari, "haplons"by H. Fritzsch, "pre-quarks" by J. D. Bjorken, etc. I wonder if Dr. Salam, who invented the terminology of "electroweak interaction" successfully, does not care whether the terminology of preon might sound like "prion" which is the origin of Jakob desease. In 1984, I proposed to call them simply "ams" in the International Workshop on Composite Models of Quarks and Leptons, Laguna Beach, California, but could not succeed in obtaining a unanimous approval from the audience, unfortunately.

2) I first met Dr. Salam when he gave a seminar on the grand unification of strong and electroweak interactions at The Rockefeller University in early nineteen- seventies. That was the period of time during which he frequently visited New York City in order to work at the United Nation Headquarters for founding the International Centre for Theoretical Physics(ICTP) in Trieste. It is amazing indeed that he has succeeded in founding the now world famous institute almost all by himself within a short period for the people in the world-wide developing countries in Africa, Asia, Oceania, North America, and South America. The ICTP has been grown up rapidly, supported not only by the United Nations but also by many developed countries (but mainly by Italy). It has only a small number of permanent staffs, but has always many visiting scientists (and also engineers, these days) coming from all over the world as it organizes many international conferences and workshops on many different fields of sciences and technologies, every year. Apparently, Dr. Salam seemed to have a desire for developing the ICTP in such a way that it may finally become the International Centre for Science (and Technology)(ICS or ICST). Unfortunately, he could not succeed in reaching that goal while he was alive. However, I suppose that every one in the developing countries hopes that the successive Directors of

the Abdus Salam International Centre for Theoretical Physics would succeed in making his dream come true!

3) I then met Dr. Salam when he came to Japan for participating in the XIX International Conference on High Energy Physics, Tokyo, 1978. I was asked to be not only the organizer of the parallel session on the grand unification of all interactions including gravity but also the scientific secretary for Dr. Salam to present a rapporteur talk in the plenary session on the same subject. Before the Conference, both the organizers of the parallel sessions and the invited speakers of rapporteur talks in the plenary sessions were requested to get together in the pre-conference held in the Institute of Fundamental Physics, Kyoto University. I was asked to pick up Dr. Salam at Kyoto Station. It was terribly hot and humid that day so that I was in a T-shirt and short pants. Therefore, I was amazed to see him coming out of the station, wearing a heavy suit, and saying, "Hi! It's cool down here!" Later, I heard that he had been born in the hottest city of Pakistan.

After a brief business meeting at the pre-conference, we took him a short sight-seeing tour in Kyoto as it seemed to be his first trip to Kyoto. In Kiyomizu-dera, one of the most popular Buddist temples in Kyoto, I told him that the "Kiyomizu-no-butai"(the stage which is extremely high above the ground) is the most famous or notorious cite for comitting suicides (by jumping over the terrace) in Japan and that there is a Japanese phrase, "Kiyomizu-no-butai kara tobioriru kimochide", which means "with the maximum bravery as if one would dare to commit a suicide". Then, he said to me, "Oh, no! I do not have to jump over there now that the $SU(2) \times U(1)$ has survived, being consistent with the recent experimental data."

4) Again I met Dr. Salam when I participated in the Second Marcel Grossmann Meeting on the Recent Developments of General Relativity, Miramare-Trieste, Italy, 1979. That international conference sponsored by the United Nations was dedicated to the hundredth anniversary of the birth of Albert Einstein and was very much exotic as there were one hundred members in the Organizing Committee or International Advisory Committee and as there were one hundred invited speakers all in the plenary session. It may be due to the fame of Dr. Salam, the organizer of the Meeting, or that of Dr. Einstein, the founder of General Relativity, that there appeared a large number of the participants including many legendary physicists such as Dr. P. A. M. Dirac. I was lucky to have been invited as one of the hundred speakers for giving a talk on the unified model of all fundamental particles and forces in front of such a decent audience although I had not yet contributed anything to General Relativity. However,

all the participants seemed to miss the only one of the hundred speakers, Dr. Zel'dovich, who could not come out of the Soviet Union, but might be satisfied with seeing his vivid picture in the video movie projected on the screen in the Auditorium of ICTP. It was also a pity for us to have very little time for seeing Dr. Salam that time since he was the busiest man in the Meeting as not only the organizer but also the Director of ICTP.

5) I met with Dr. Salam when he came to Japan for a business meeting at the United Nation University (UNU) Headquarters in Tokyo as he was the Rector of UNU. It was in 1980, the only one year after he had received Nobel Prize in physics together with S. L. Glashow and S. Weinerg for their construction of the unified $SU(2) \times U(1)$ gauge theory of electroweak interactions. Before he came to Japan, he had sent me a letter in which he asked me to help him in obtaining more financial support to the ICTP from our Japanese government. After his arrival in Tokyo, we first went to the Headquarter of the leading party in the Congress of Japan where we had an appointment to see the No. 2 in the party. After waiting so long, Dr. Salam finally met the man for a few minutes or so and asked him a little help. Later, Dr. Salam told me that it had worked.

The next day, I was asked to be an observor in the top administration meeting of the UNU. Dr. Salam introduced me to the president of UNU as his most powerful helper in Japan. For the most of time in the meeting, they were discussing how they could receive more financial supports from the developed countries in the western worlds. When we met incidentally the well-known French theoretical physicist, who was also working there, he said to me, "What are you doing here?", I replied, "I am a student of Dr. Salam". Then, Dr. Salam instantly denied by saying to him, "No! No! He is my friend!".

The day after the next day, Dr. Salam gave us a colloquium on the super-grand unification of all fundamental interactions including gravity at Institute for Nuclear Study, University of Tokyo. After the colloquium, we made a press conference in front of the reporters of the three main newspapers in Japan. He emphasized that the ICTP needs more financial support from the Japanese government. The only one of the three newspapers, The Asahi Shimbun, put a large article on his request, but the other two seemed to ignore it. After the press conference, we moved to The Embasy of Pakistan for the big celebration party for the Nobel Laureate where the Ambassador of Pakistan in Japan and many Japanese physicists invited were waiting for us.

6) In July-August of 1981, I was invited by Dr. Salam to visit (together with my family) the International Centre for Theoretical Physics (ICTP) as a visiting

Professor for a month. I rent an appartment in the downtown area of Trieste, and was commuting between the Down Town and the ICTP by changing local buses twice in front of Central Station of Trieste and at Barcola. In the Centre, I was provided with a very cosy office and with a kind help of the secretary so that I may have finished writing the paper entitled, "Pregeometric Origin of the Big Bang", for which Professor Keiichi Akama, the co-author and my life-long collaborator, and I had received the Gravity Research Foundation Award in 1982! I really appreciated the courtecy service from the Director and his secretaries for producing hundreds of preprints of the paper and for distributing them to hundreds of institutions and universities all over the world.

It was also a wonderful experience for me to be associated with Dr. Salam every days for a month, especially during lunch time at the cafeteria in the Centre. It would recall me the old happy days when I had lunch almost every days in 1970-1971 together with Professor Hans Bethe and the others at the cafeteria of Hilton in Cornell University and also in 1971-1975 together with Professor Abraham Pais and the others at the one in The Rockefeller University. Only once, I was refused to join him at the table which had already been fully occupied by African and Middle-Asian physicists on the day when the Islamic sacred period was just over. That was the only one occasion when I felt a gap between him and me.

7) In June-July, 1982 I was visiting the Max Planck Institute in Munchen as a visiting Research Fellow, preparing for a rappeutor talk on the super-high energy physics in the XXI International Conference on High Energy Physics, Paris, 1982. Then, I was suddenly called up by Dr. Salam and requested to visit the International Centre for Theoretical Physics as a visiting Professor again, but only for a week. The main purpose for him to have invited me at that time seemed to be discussing how to receive more financial supports from the Japanese government. He asked me a lot of simple questions such as "What is the mail address of the Headquarters of the leading party in the Japanese Congress?". I gave him many simple answers such as "Headquarters of Free Democratic Party, Tokyo, Japan is enough". I wondered if such a simple help from me would have deserved the very large amount of stipend and travel expense which I had received from the ICTP for my short visit from Munchen. Thanks to his invitation, I enjoyed mountain-climbing in the Austrian Alps and in the Dolomite Alps on my way to Trieste while I did sight-seeing in Wien and in Salzburg on my way back to Munchen.

8) In 1988, when I was participating the Fifth Marcel Grossmann Meeting on Recent Developments in Theoretical and Experimental General Relativity, Gravitation and Relativistic Field Theories, Perth, 1988, I met Professor Mainuddin Ahmed, a mathematician and general-relativity-theorist from Bangladesh in the train on which we were riding for the excursion during the Meeting. He seemed to be a frequent visitor to the ICTP and was worrying about the future of the ICTP after Dr. Salam, the Director of ICTP at that moment, would die. He also strongly insisted that we should build another international center for theoretical physics similar to the ICTP somewhere in Asia. I knew that Dr. Salam was not only mentally but also physically very strong so that the Director of ICTP (who was 61 or 62 at that moment) might live long enough to develope the ICTP into the International Centre for Science (and Technology) within a decade or so. Therefore, after I had been back to Tokyo from Perth, I made up the CAOS Project for building the Center of Asia and Oceania for Science (CAOS), to be located somewhere in Asia or Oceania and to be completed by 2001. On April 1, 1989, Professor Mainuddin Ahmed (as the Director of the Center in Bangladesh), Professor Yuichi Chikashige who was the one of my collaborators (as the Vice-Director of CAOS), and I (as the Director-in-general of CAOS) founded the CAOS only with the Headquartors in Tokyo, the Observatory of COSMOS(Center of Oizumi for Science and Mansion of Oizumi for Scientists) in Yamanashi, Japan, and the Center of Bangladesh in Rajshahi, Bangladesh.

9) Then in June, 1989, I last met Dr. Salam by participating in the First BCSPIN Summer School on Physics, Kathmandu, Nepal, which was organized for educating graduate students and young physicists from some Asian developing countries of BCSPIN(Bangladesh, China, Sri Lanka, Pakistan, India, and Nepal) by several teachers from the European countries and the United States. I was the only one participant invited as a sort of observer from the Asian developed country of Japan. I arrived at Kathmandu on June 6, the National holiday for celebration of the birth of Budda. The next day, I saw Dr. Salam having a cane and heard that he had been suffering from the Parkinson disease. In the Opening Ceremony or Banquet of the School, all the participants were invited to be in the Royal Palace and to see the King of Nepal who was later tragically killed by the Prince.

Much earlier than the end of the BCSPIN School, I must leave the lovely city of Kathmandu in order to present the colloquium on pregeometric theory of gravitation and to discuss on how to develop the CAOS Project in Rajshahi

University, Bangladesh. From the Royal Nepal Airline plane flying from Kathmandu to Dacca, I saw both the highest mountain in the world and the most horrible flood in Bengal and Bangladesh. The next day, I arrived at Rajshahi by a small plane of the domestic airline of Bangladesh from Dacca, met Professor Mainuddin Ahmed at the Airport, and found the city even more beautiful than Kathmandu and suitable for the location of the Center of Bangladesh for Science!

10) My wife and I tried to meet Dr. Salam in July-August 1995 by visiting the International Centre for Theoretical Physics, but it was in vain. Right after I arrived at the ICTP, I was told by the secretary that he was seriously ill at his home in London so that he could not come to Trieste any more. Without having him, the ICTP looked like a dry oasis for me coming from a developed country although it was running well for physicists from developing countries, under the control of Dr. Virasoro, the new young Director. In fact, I was lucky to have seen by chance the Bangladesh friend, who was participating in a workshop going on at that moment there in the ICTP. As he must rash back home, I could not have enough time to disscuss how to develop the CAOS Project. I also missed seeing him in the clouded seminar room when I gave a seminar on the New Large Number Hypotheses in the Universe in order to compensate a part of the local expenses which I had received from the ICTP.

A year had not passed when, unfortunately, Dr. Salam died at the age of 69 or 70, in 1996. He has passed away without seeing his dream of developing the ICTP into the International Center for Science (and Technology) come true. Furthermore, my only friend in Bangladesh also passed away without coming to Tokyo to see me at the Headquarters of CAOS. However, I still have a dream of developing the CAOS into the Asian and Oceaniac Branch of the Abdus Salam International Center for Science (and Technology)!

11) There were many reasons why we could not quickly develop the CAOS Project since we founded the CAOS on April 1, 1989. First of all, as I was an associate professor in Institute for Nuclear Study, University of Tokyo, a National servant of Japan, I must obtain a permission from Minister of Education, Science, Sports and Culture of Japan in order to work simultaneously as Director of the CAOS. In order to obtain the permission, I must submit the required documents together with the recommendation letter from the Director in Bangladesh to the Director of Institute for Nuclear Study(INS). Furthermore, in order to request the President of University of Tokyo to ask the Ministry to give me the permission, the Director of INS must ask a unanimous approval in the

faculty meeting of INS including me as a member. I have been waiting for the last almost three decades to receive the permission, and still do not understand why the Director of INS had never asked the faculty meeting on my request. By the way, the Institute of Nuclear Study, University of Tokyo was closed in 1997 and have been united together with the Meson Factory, Faculty of Science, University of Tokyo, and with the National Laboratory for High Energy Physics, and restructured into the Institute of Elementary Particle and Nuclear Physics, High Energy Accelerator Research Organization.

The second reason which is more important is the following: the world has been changing very rapidly for the last three decades. The cold war has ended and the IT has advanced so much that anyone in a country may now communicate freely and instantly with another one in another country by SNS. There seems to be little necessity any more for people to get together at a certain place on this planet Earth in order to discuss on anything like science. Therefore, there seems to be no necessity for us, scientists, to make efforts for constructing "hard or solid" institutes for "soft sciences" including theoretical physics. Therefore, I have decided to make the CAOS as soft as possible, something like a network of institutes for science in Asia and Oceania.

12) In 2003, I was asked by Professor Mainuddin Ahmed to be Advisor to Midlands Academy of Business and Technology (MABT) which his son, Mr. Rokonuddin Ahmed, was about to found in Leicester, United Kingdom. As I was a staff in the Institute for Elementary Particle and Nuclear Study, High Energy Accelerator Research Organization, in order to accept the offer for becoming a staff in the organization in a foreign country, I must obtain a permission from the Prime Minister of Japan!

In September, after I attended a international conference on physics in Minsk, I went to Leicester in order to see the son of my friend and what the status of MABT was. He picked me up by his car at the airport in London and took me to the beautiful city in Midlands. I met his family at his home and his colleagues including a lady English teacher and a young IT scientist at the small business office in the community center. There were the only three staffs and, of course, no student in the MABT before its foundation as in the CAOS on its foundation in 1989!

On my way back to London by his car, he asked me to become the founding president of MABT. I instantly declined to accept the offer as I knew it illegal for me, a civil servant of Japan, to become a head of any organization in any foreign country.

One and a half decades have passed since the foundation of MABT, I am now very much delighted to know that he, not only as the CEO of MABT but also as the Principal of MABT, is smart enough to have developed it so quickly that it has now a few tens of staff and hundreds of students. I remembered to hear that Dr. Salam had developed the ICTP quickly almost all by himself, starting at a single room in the University of Trieste on its foundation!

Chapter II

Composites of Subquarks as Quark Matter*

Abstract

Not only the masses of fundamental particles including the quarks and leptons, and the weak and Higgs bosons, but also the mixing angles of quarks and those of neutrinos are all explained and/or predicted in the unified composite model of all fundamental particles and forces successfully.

This Chapter consists of the following Sections: I. Introduction, II. Unified Composite Model, III. Quark Mixing Angles, IV. Quark and Charged Lepton Masses, V. Weak and Higgs Boson Masses, VI. Future Prospects, and Appendix–Neutrino Masses and Mixing Angles.

It also contains References of and Dictionary for Chapter II. Composites of Subquarks as Quark Matter.

Between Chapter II and Chapter III, Picture of Dr. Andrei Sakharov and the Author, and Episodes on Dr. Andrei Sakharov and the Author are added as Dr.

---

*Some contents of this Chapter have been presented in the contributed paper entitled, "Mass Spectrum of Quarks and Leptons in the Unified Composite Model", to the XXIV International Seminar, "Nonlinear Phenomena in Complex Systems", Minsk, 2017, published in the international journal of Nonlinear Phenomena in Complex Systems **20** (2017)327, which is an extended and updated version of the previous contributed paper entitled, "Masses of Fundamental Particles (*Dedicated to the late Professor Kazuhiko Nishijima*)", published in the Proceedings of the XXII International Conference on New Trends in High-Energy Physics, Alushta (Crimea), 2011, edited by P. N. Bogolyubov and L. L. Jenkovszky (Bogolyubov Institute for Theoretical Physics, Kiev, 2011), p.352, arXiv:1109.3705v5[physics.gen-ph] 10 Feb 2012.

Sakharov was the founder of Pregeometry, to be discussed in Chapter III.

I. Introduction

In 1953, Nakano and Nishijima, and independently Gell-Mann [1] found "strangeness", the new quantum number for particles, which is the beginning of hadron physics in the third quarter of the twentieth century. Then, Gell-Mann and independently Zweig [2] introduced the **quark model for hadrons**, and Nambu and the others [3] proposed QCD(quantum chromodynamics), the **Yang-Mills gauge theory** of color $SU(3)$ for strong interactions of quarks and gluons. Furthermore, Glashow, Salam, and Weinberg [4] proposed **QFD(quantum flavor dynamics)**, the $SU(2) \times U(1)$ gauge theory for electroweak interactions of quarks and leptons. Thus, the **Standard Model** of all elementary particle forces consisting of QCD and QFD, had already been not only proposed theoretically but also confirmed experimentally by the discoveries of the weak and Higgs bosons. Therefore, what is left in quark-lepton physics is to explain and/or to predict not only all the masses of fundamental particles including the quarks and leptons, and the weak and Higgs bosons, but also all the mixing angles of quarks [5] and those of neutrinos [6]. A main purpose of this Chapter is to discuss how to explain and/or to predict all of these fundamental parameters in the Standard Model successfully in the unified composite model of quarks and leptons [7, 8].

II. Unified Composite Model

In 1977, we proposed the unified composite model of fundamental particles and forces in which not only quarks and leptons but also gauge bosons and Higgs scalars are composites of subquarks, the most fundamental form of matter [7, 8]. The minimal supersymmetric composite model of quarks and leptons consists of an isodoublet of spinor subquarks with charges $\pm 1/2$, w_1 and w_2(called "wakems" standing for **weak** and **electromagnetic**), and a Pati-Salam color-quartet of scalar subquarks with charges $+e/2$ and $-e/6$, C_0 and $C_i (i = 1,2,3)$(called "chroms" standing for colors). The spinor and scalar subquarks with the same charge $+e/2$, w_1 and C_0, may form a fundamental multiplet of $N = 1$ **supersymmetry** [9]. Also, all the six subquarks, $w_i (i = 1,2)$ and $C_\alpha (\alpha = 0,1,2,3)$, may have "subcolors", the additional degrees of freedom, and belong to a fundamental representation of subcolor symmetry [10]. Although the subcolor symmetry is unknown, a simplest and most likely candidate for it is $SU(4)$. Therefore, for simplicity, all the subquarks are assumed to be quartet in subcolor $SU(4)$. Also, although the confining force is unknown, a simplest and most likely candidate for it is the one described by quantum subchromodynamics(QSCD), the Yang-Mills gauge theory of subcolor $SU(4)$ [10]. Note that the subquark charges satisfy not only the **Nishijima- Gell-Mann rule** of $Q = I_{w3} + (B-L)/2$ but also the "**anomaly- free condition**" of $\sum_i Q_{w_i} = \sum_\alpha Q_{C_\alpha} = 0$. The Lagrangian for QSCD is simply given by

$$L_{QSCD} = \sum_i \overline{w}_i[i\gamma^\mu(\partial_\mu - ig\frac{\lambda^a}{2}A^a_\mu) - M_i]w_i$$

$$+\sum_\alpha[|(\partial_\mu - ig\frac{\lambda^a}{2}A^a_\mu)C_\alpha|^2 + m^2_\alpha|C_\alpha|^2]$$

$$-\frac{1}{4}(\partial_\mu A^a_\nu - \partial_\nu A^a_\mu + gf^{abc}A^b_\mu A^c_\nu)^2,$$

where g is the coupling constant of $SU(4)_{sc}$ gauge fields, $A^a_\mu(a = 1,2,3...,15)$, λ^a and f^{abc} are the extended **Gell-Mann matrices** and structure constants of $SU(4)$, respectively, and M's and m's are the wakem and chrom masses.

In the minimal supersymmetric composite model, we expect that there exist at least $36(=6\times6)$ composite states of a subquark and an antisubquark which are subcolor-singlet. They include 1) $16(=4\times2\times2)$ spinor states corresponding to one generation of quarks and leptons, and their antiparticles of

$$\nu = \overline{C}_0 w_1, l = \overline{C}_0 w_2, u_i = \overline{C}_i w_1, d_i = \overline{C}_i w_2,$$

and the h.c.'s, and 2) $4(=2\times2)$ vector states corresponding to the photon and weak vector bosons (or 4 scalar states corresponding to the Higgs scalars) of

$$V_{ij}(\text{or } \phi_{ij}) = \begin{pmatrix} \overline{w}_1 w_1 & \overline{w}_2 w_1 \\ \overline{w}_1 w_2 & \overline{w}_2 w_2 \end{pmatrix}$$

$(i,j = 1,2)$, and 3) $16(=4\times4)$ vector states corresponding to a) the gluons, "leptogluon", and "barygluon" of

$$G^a = \overline{C}_i \frac{\lambda^a_{ij}}{2} C_j; G^0 = \overline{C}_0 C_0; G^9 = \overline{C}_i C_i$$

$(i, j = 1, 2, 3)$, where $\lambda^a (a = 1, 2, 3, ..., 8)$ is the Gell-Mann matrix of $SU(3)_c$, and b) the "vector leptoquarks" of

$$X_i = \overline{C}_0 C_i$$

and the hermitian conjugates$(i = 1, 2, 3)$, or $16(= 4 \times 4)$ scalar states corresponding to the "scalar gluons", "scalar leptogluon", "scalar barygluon", and "scalar leptoquarks" of

$$\Phi_{\alpha\beta} = \overline{C}_\alpha C_\beta$$

$(\alpha, \beta = 0, 1, 2, 3)$. Quarks and leptons with the same quantum numbers but in different generations can be taken as dynamically different composite states of the same constituents. In addition to these "meson-like composite states" of a subquark and an antisubquark, there may also exist "baryon-like composite states" of 4 subquarks which are subcolor-singlet. These exotic quarks, leptons, and gauge bosons are new forms of quark-lepton matter, which is a subject to discuss later in Section IV.

In the unified subquark model of quarks and leptons [8], it is an elementary exercise to derive the **Georgi-Glashow relations** [11],

$$sin^2\theta_w = \sum(I_3)^2 / \sum Q^2 = 3/8$$

and

$$f^2/g^2 = \sum(I_3)^2 / \sum(\lambda^a/2)^2 = 1$$

for the **weak-mixing angle**(θ_w), the gluon and weak-boson coupling constants (f and g), the third component of the isospin (I_3), the charge (Q), and the color-spin ($\lambda^a/2$) of subquarks, without depending on the assumption of grand unification of strong and electroweak interactions. The experimental value is $sin^2\theta_w(M_Z) = 0.23129(5)$ [12]. The disagreement between the value of $3/8$ predicted in the subquark model and the experimental value might be excused by insisting that the predicted value is viable as the running value renormalized *a la* Georgi, Quinn, and Weinberg at extremely high energies (as high as $10^{15} GeV$), given the "**desert hypothesis**"[13].

III. Quark Mixing Angles

In the unified composite model of quarks and leptons [14], the **quark-mixing matrix** V [5] is given by the expectation value of the subquark current between the up and down composite states as

$$V_{us} \sim < u|\overline{w}_1 w_2|d >, \ldots$$

[15]. By using the **algebra of subquark currents** [16], the unitarity of quark-mixing matrix, $VV^+ = V^+V = 1$, have been demonstrated although the superficial non-unitarity of V as a possible evidence for the substructure of quarks has also been discussed by myself [17]. In the first-order perturbation of isospin breaking, we have derived the relations of

$$V_{us} = -V_{cd}^*, V_{cb} = -V_{ts}^*, \ldots$$

which agree well with the experimental values of $|V_{us}| = 0.2248 \pm 0.0006$ *vs.* $|V_{cd}| = 0.220 \pm 0.005$ and $|V_{cb}| = 0.0405 \pm 0.0015$ *vs.* $|V_{ts}| = 0.0400 \pm 0.0027$ [12], and some other relations such as

$$|V_{cb}|(= |V_{ts}|) \cong (m_s/m_b)V_{us} \cong 0.021,$$

which roughly agree with the recent experimental value of $V_{cb} = 0.0405 \pm 0.015$ [12]. In the second-order perturbation, the relations of

$$|V_{ub}| \cong (m_s/m_c)|V_{us}V_{cb}| \cong 0.0017$$

and

$$|V_{td}| \cong |V_{us}V_{cb}| \cong 0.0046$$

have been predicted. The former relation agrees roughly with the recent experimental data $|V_{ub}| = 0.00409 \pm 0.00039$. The predictions for V_{ts} and V_{td} also agree fairly well with the experimental estimates from the assumed unitarity of V, $|V_{ts}| = 0.0400 \pm 0.0027$ and $|V_{td}| = 0.0082 \pm 0.0006$ [12]. To sum up, we have succeeded in predicting all the magnitudes of the **CKM matrix** elements except for a single element, say, V_{us}.

IV. Quark and Charged Lepton Masses

For the last quarter century, we have been trying to complete an ambitious program for explaining all the quark and lepton masses by deriving many sum rules and/or relations among them and by solving a complete set of the sum rules and relations [18]. By taking the first generation of quarks and leptons as almost **Nambu-Goldstone fermions** due to spontaneous breakdown of approximate supersymmetry between a wakem and a chrom [19], and the second generation of them as quasi Nambu-Goldstone fermions, the superpartners of the **Nambu-Goldstone bosons** due to spontaneous breakdown of approximate global symmetry [20], we have not only explained the hierarchy of quark and lepton masses,

$$m_e \ll m_\mu \ll m_\tau, m_u \ll m_c \ll m_t, m_d \ll m_s \ll m_b,$$

but also obtained the square-root sum rules for quark and lepton masses [21],

$$m_e^{1/2} = m_d^{1/2} - m_u^{1/2}$$

and

$$m_\mu^{1/2} - m_e^{1/2} = m_s^{1/2} - m_d^{1/2},$$

and the simple relations among quark and lepton masses [22],

$$m_e m_\tau^2 = m_\mu^3$$

and

$$m_u m_s^3 m_t^2 = m_d m_c^3 m_b^2,$$

all of which are remarkably well satisfied by the experimental values and estimates. By solving a set of these two sum rules and two relations, we can obtain the following predictions:

$$\begin{pmatrix} m_e & m_\mu & m_\tau \\ m_u & m_c & m_t \\ m_d & m_s & m_b \end{pmatrix} =$$

$$
\begin{pmatrix}
0.511MeV_{(input)} & & \\
& 105.7MeV_{(input)} & \\
& & 1520MeV_{(1776.86\pm0.12MeV)} \\
2.2\begin{matrix}+0.6\\-0.4\end{matrix}MeV_{(input)} & & \\
& 1.27\pm0.03GeV_{(input)} & \\
& & 171\pm17GeV_{(173.21\pm0.51\pm0.71GeV)} \\
4.8\begin{matrix}+0.9\\-0.6\end{matrix}MeV_{(4.7\begin{matrix}+0.5\\-0.4\end{matrix}MeV)} & & \\
& 138\pm2MeV_{(96\begin{matrix}+8\\-4\end{matrix}MeV)} & \\
& & 4.18\begin{matrix}+0.04\\-0.03\end{matrix}GeV_{(input)}
\end{pmatrix}
$$

where the "inputs"and the values indicated in the parentheses denote either the experimental data [12] or the phenomenological estimates [12, 23], to which our predicted values should be compared. Furthermore, if we solve a set of these two sum rules and two relations, and the other two sum rules obtained in the unified composite **model of the Nambu- Jona-Lasinio type** [24] for all fundamental particles and forces [15],

$$m_W = (3 < m_{q,l}^2 >)^{1/2}$$

and

$$m_H = 2(\sum m_{q,l}^4 / \sum m_{q,l}^2)^{1/2},$$

where $m_{q,l}$'s are the quark and lepton masses and $<>$ denotes the average value for all the quarks and leptons, we can predict not only the four quark and/or lepton masses such as m_d, m_s, m_t, and m_τ as above but also the Higgs scalar and weak boson masses as

$$m_H \cong 2m_t = 342 \pm 34GeV$$

and

$$m_W \cong \sqrt{3/8}m_t = 105 \pm 10GeV,$$

which should be compared to the experimental value of $m_W = 80.385 \pm$ 0.015GeV [12]. Recently, I have been puzzled by the "new Nambu's empirical quark-mass formula" of $M = 2^n M_0$ with his assignment of $n = 0, 1, 5, 8, 10, 15$

for u, d, s, c, b, t [25], which makes my relation of $m_u m_s^3 m_t^2 = m_d m_c^3 m_b^2$ exactly hold. More recently, I have been more puzzled by the relations of $m_u m_b \cong m_s^2$ and $m_d m_t \cong m_c^2$ suggested by Davidson, Schwartz, and Wali(D-S-W) [26], which can coexist with my relation and which are exactly satisfied by the Nambu's assignment. If we add the D-S-W relations to a set of our two sum rules, two relations, and sum rule for m_W and if we solve a set of these seven equations by taking the experimental values of $m_e = 0.511 MeV$, $m_\mu = 105.7 MeV$, and $m_W = 80.4 GeV$ as inputs, we can find the quark and lepton masses of

$$\begin{pmatrix} 0.511MeV_{(input)} & & \\ & 105.7MeV_{(input)} & \\ & & 1520MeV_{(1776.86\pm 0.12MeV)} \\ 3.8MeV_{(2.2 {+0.6 \atop -0.4} MeV)} & & \\ & 970MeV_{(1.27\pm 0.03GeV)} & \\ & & 131.3\pm 0.2GeV_{(173.21\pm 0.51\pm 0.71GeV)} \\ 7.2MeV_{(4.7 {+0.5 \atop -0.4} MeV)} & & \\ & 150MeV_{(96 {+8 \atop -4} MeV)} & \\ & & 5.9GeV_{(4.18 {+0.04 \atop -0.03} GeV)} \end{pmatrix}$$

where an agreement between the calculated values and the experimental data or the phenomenological estimates [12, 23] looks reasonable. This result may be taken as one of the most elaborated "modern developments in elementary particle physics". To sum up, we have succeeded in explaining and predicting most of the properties (masses and mixing angles) of quarks and leptons in the unified supersymmetric composite model.

Now I am ready to discuss new forms of quark-leptonic matter such as exotic quarks, leptons, and gauge bosons. Among "meson-like states" of subquarks, not only the leptogluon $G^0 (= \overline{C}^0 C^0)$, barygluon $G^9 (= \overline{C}_i C_i)$, and vector leptoquarks $X_i (= \overline{C}_0 C_i)$, but also the scalar gluons, scalar leptogluon, scalar barygluon, and scalar leptoquarks $\Phi_{\alpha\beta} (= \overline{C}_\alpha C_\beta)$ are all exotic since they are not familiar fundamental particles accommodated in the Standard Model. The masses of these exotic gauge bosons may be generated by spontaneous breakdown of the $SU(4)_c$ symmetry. In the unified composite model of quark-lepton color interactions proposed by myself [27], the $SU(4)_c$ transmuted from $SU(4)_{sc}$ [28] is spontaneously broken down to $SU(3)_c$ due to the condensation of $\overline{C}_0 C_0 (< \overline{C}_0 C_0 >= (V/\sqrt{2})^2 \neq 0)$ and the B-L gluon $G^{B-L}[=\overline{C}_i(\lambda^{15}/2)_{ij}C_j]$ becomes $\sqrt{3/2}$ times heavier than the vector leptoquarks X_i(whose mass is given by $fV/2$ where f is the gluon coupling constant). Since the gluon coupling constant f is determined by $f = (4\pi\alpha_s)^{1/2} \cong 1.2$(for the renormalization point at m_Z^2), the vacuum expectation value V can be bounded as $V = m_X/f \geq 0.5TeV$. This lower bound on V seems to be consistent with the expectation that the energy scale of QSCD, Λ_{sc}, must be much higher, say, $\Lambda_{SC} \geq 1TeV$, than that of QCD, $\Lambda_C(\cong 0.2GeV)$. In short, the physical B-L gluon should appear as a neutral, color-singlet, but strongly interacting vector boson which is $\sqrt{3/2}$ times heavier than the vector leptoquarks and which couples universally with the B-L current of quarks and leptons. This mixture of leptogluon and barygluon, if exists, may look like the source of every matter as it decays strongly into all matter-antimatter pairs. In any case, we expect that all these exotic gauge bosons and Higgs scalars must be much heavier than the ordinary weak bosons ($W^{\pm}$ and Z) and Higgs scalars(H's).

The most exotic is the scalar color-ball of $\varepsilon_{ABCD}\varepsilon_{\alpha\beta\gamma\delta}\ C^A_\alpha\ C^B_\beta\ C^C_\gamma\ C^D_\delta$. It is not only neutral electromagnetically but also neutral weakly, but it has strong interactions with any hadrons due to the van der Waals force of the $(C_1C_2C_3)$ content as a baryon, a color singlet state of three quarks $(q_1q_2q_3)$, has strong interactions with any other hadrons. It may be called "primitive hydrogen" since it has both the baryon and lepton numbers. It can be taken another source of the baryon and lepton number asymmetry in the Universe. Also, since its mass may be as large as $\Lambda_{SC}(\geq 1TeV)$, since its size may be as small as $1/\Lambda_{SC}(\leq 1/1TeV)$, and since it may be absolutely stable because of baryon and lepton number conservations, it can be another good candidate for a particle responsible for the missing mass in the Universe.

V. Weak and Higgs Boson Masses

In the unified $SU(2)\times U(1)$ gauge theory of electroweak interactions [4], the masses of weak bosons, $W^{\pm}$ and Z, are given as

$$m_W = m_Z cos\theta_w = (\pi\alpha/\sqrt{2}G_F)^{1/2}/sin\theta_w$$

(where α and G_F are the fine structure and Fermi coupling constants) so that they can be predicted if the weak mixing angle θ_w is fixed by any means. In 1969, the weak boson mass was predicted for the first time to be of the order of 100GeV by myself, provided that the self-masses of leptons be finite [29]. In the grand unified $SU(5)$ gauge theory of strong and electroweak interactions of Georgi and Glasgow [11] or in the unified composite model of quarks and leptons [14], the weak angle is fixed as $sin^2\theta = 3/8$ so that the weak boson masses, m_W and m_Z, may be predicted.

In 1980, the mass of the physical Higgs boson in QFD, m_H, was first predicted by myself in the composite model of Higgs bosons [30]. To be more precise, it has been predicted in the following three ways:

In general, in composite models of the Nambu-Jona-Lasinio type [24], the Higgs boson appears as a composite state of fermion-antifermion pairs with the mass twice as much as the fermion mass. Our unified subquark model of the Nambu-Jona-Lasinio type [14] has predicted the following two sum rules:

$$m_W = [3(m_{w_1}^2 + m_{w_2}^2)/2]^{1/2}$$

and

$$m_H = 2[(m_{w_1}^4 + m_{w_2}^4)/(m_{w_1}^2 + m_{w_2}^2]^{1/2},$$

where m_{w_1} and m_{w_2} are the wakem masses while m_W and m_H are the charged weak boson and physical Higgs boson masses in the Standard Model, respectively. By combining these sum rules, the following relation has been obtained for $m_{w_1} = m_{w_2}$:

$$m_w : m_W : m_H = 1 : \sqrt{3} : 2.$$

From this relation, the wakem and Higgs boson masses have been predicted as

$$m_w = m_W/\sqrt{3} = 46.4GeV$$

and

$$m_H = 2m_W/\sqrt{3} = 92.8GeV$$

for $m_W = 80.4GeV$. More precisely, from the two sum rules, the Higgs boson mass can be bounded as

$$92.8GeV = 2m_W/\sqrt{3} \leq m_H \leq 2\sqrt{6}m_W/3 = 131.3GeV.$$

On the other hand, our unified quark-lepton model of the Nambu-Jona-Lasinio type [14] has predicted the two sum rules for m_W and m_H presented in the last Section. If there exist only three generations of quarks and leptons, these sum rules completely determine the top quark and Higgs boson masses [30] as

$$m_t \cong (2\sqrt{6}/3)m_W = 131.3GeV$$

and

$$m_H \cong 2m_t \cong (4\sqrt{6}/3)m_W = 262.6GeV.$$

Furthermore, trinity of hadrons, quarks, and subquarks [31] tells us that these sum rules can be further extended to the approximate sum rules of

$$m_W \cong (3 < m_{B,l}^2 >)^{1/2}$$

and

$$m_H \cong 2(\sum m_{B,l}^4 / \sum m_{G,l}^2)^{1/2},$$

where $m_{B,l}$'s are the "canonical baryon" and lepton masses and $<>$ denotes the average value for all the canonical baryons and leptons. The "canonical baryon" means either one of p, n and other ground-state baryons of spin 1/2 and weak-isospin 1/2 consisting of a quark heavier than the u and d quarks and a scalar and isoscalar diquark made of u and d quarks. If there exist only three generations of quarks and leptons, these sum rules completely determine the masses of the canonical topped baryon, T, and the Higgs scalar as

$$m_T \cong 2m_W = 160.8GeV$$

and

$$m_H \cong 2m_T \cong 4m_W = 321.6GeV.$$

Therefore, since the Higgs boson mass has been found to be $m_H = 125.09 \pm 0.24$GeV [12] between 93GeV and 131GeV, it looks like a composite state of subquark-antisubquark pairs. Furthermore, the two sum rules in the unified subquark model of the Nambu-Jona-Lasinio type completely determine the subquark masses as (m_{w_1}, m_{w_2}) =($13GeV$, $64GeV$) or ($64GeV$, $13GeV$) for $m_H = 125GeV$ and $m_W = 80.4GeV$. It suggests that either one of w_1 and w_2 vanishes or is much smaller than the other.

VI. Future Prospects

What can we expect to find in future high-energy experiments and astroparticle observations.

First of all, many of us may expect that superparticles [9], the superpartners of fundamental particles, will be found if the energy scale of SUSY breaking is of the order of 1 TeV. In the unified supersymmetric composite model, there may exist superpartners of any fundamental particles containing w_1 or C_0, which can be made by replacing either w_1 by C_0 or C_0 by w_1. However, there is no reason for expecting that any superparticles be as light as hundreds GeV. All of them must be as heavy as the compositeness energy scale, say 1TeV.

Second, some of us (in particular, string theorists) may expect that Kaluza-Klein extra dimensions of the space-time [32] would be found if the size of extra dimensions is as small as of the order of 1 TeV^{-1}, which has recently been emphasized by Arkani-Hamed et al. [33]. However, it seems to me that there is no compelling reason for expecting that the size of extra dimensions is so large as the LHC energy is enough high to see an evidence for the extra-dimensions, if any.

Third, some of us (in particular, composite modelists) may expect that the substructure of fundamental particles would be found if the energy scale of compositeness is of the order of 1TeV. It would be the most exciting to find an evidence for the substructure in search for anomalous events due to excited quarks, leptons, gauge bosons, and Higgs scalars [34], or due to exotic particles such as the barygluon, leptoquark, and color-ball discussed in the Section IV. In fact, every time when new colliders reached new energy scales in the last decade in the last century, some anomalies had been reported for possible evidence for the substructure at LEP [35], Tevatron [36], and HERA [37], although those anomalies have either disappeared or not been confirmed by later experiments. Recently, however, the ATLAS and CMS Collaborations at the LHC had reported two new anomalies: 1) the excess of di-boson resonance events recently found by the ATLAS Collaboration at about 2TeV and 2) the excess of di-photon resonance events lately reported by the ATLAS and CMS Collaborations at 750GeV [38]. The former might have been taken as an evidence for the excited state of either one of the weak and Higgs bosons, and the glueballs while the latter as another evidence for the substructure of di-boson resonance whose masses are of the order of 1TeV [39]. Also, very lately, the ATLAS Collaboration reported an experimental anomaly in WH or ZH production with an apparent excess at around 3TeV resonance mass region [40], which might be taken as another evidence for the excited state of either one of the weak and Higgs bosons!

About a quarter century has past since W and Z were discovered for evidence for the Standard Model. I hope that LHC and ILC will provide us enough high energy and luminosity to lead us to subquark physics in the first quarter of the 21st century [41]. The time has come when we are about to be able to give an answer to both of the following questions which Einstein seemed to ask by himself: "What is the Universe?" and "What is the electron?" [42].

Appendix-Neutrino Masses and Mixing Angles [43]

In 1997, the Super-Kamiokande Collaboration [44] found an evidence for the non-vanishing mass for the muon and/or tau neutrinos in the analysis based on **neutrino oscillation** due to the **neutrino mixing** among three generations of neutrinos [6] $(\nu_e, \nu_\mu, \nu_\tau)$, breaking **lepton-number conservation** which Nishijima, Schwinger, and Bludman had introduced four decades earlier in 1957 [45]. The existing neutrino oscillation data [12] have given rather stringent limit on the neutrino mass-squared differences and mixing angles:

$$
\begin{aligned}
&\Delta m_{21}^2 \cong 7.37(6.93-7.97)\times 10^{-5} eV^2,\\
&sin^2\theta_{12} \cong 0.297(0.250-0.354),\\
&|\Delta m^2 (= \Delta m_{31}^2 - \Delta m_{21}^2/2)|\\
&\qquad \cong \begin{cases} 2.50(2.37-2.63)\times 10^{-3} eV^2 \text{ or} \\ 2.46(2.33-2.60)\times 10^{-3} eV^2 \end{cases},\\
&sin^2\theta_{23} \cong \begin{cases} 0.437(0.379-0.616) \text{ for positive } \Delta m^2 \\ 0.569(0.383-0.637) \text{ for negative } \Delta m^2 \end{cases},\\
&sin^2\theta_{13} \cong \begin{cases} 0.0214(0.0185-0.0246) \text{ for positive } \Delta m^2 \\ 0.0218(0.0186-0.0248) \text{ for negative } \Delta m^2 \end{cases},\\
&\delta/\pi \cong \begin{cases} 1.35(0.93-1.99) \text{ or} \\ 1.32(0.83-1.99) \end{cases},
\end{aligned}
$$

The extremely small mass difference and almost maximal mixing between two neutral particles remind us of those between K^0 and $\overline{K}^0$. In analogy to the $K^0 - \overline{K}^0$ mixing, suppose that the three neutrinos $(\nu_e, \nu_\mu, \nu_\tau)$ are originally massless and without any mixings but that they have transitions between themselves due to some unknown mechanism. Then, the neutrino mass matrix has the form of

$$\begin{pmatrix} 0 & \mu & m \\ \mu & 0 & M \\ m & M & 0 \end{pmatrix}$$

where m, μ, and M are unknown parameters for the transition matrix elements. Suppose that $m \ll \mu \ll M$, and then $sin^2\theta_{23} \cong 0.5$, which is consistent with the experimental data [12]. In fact, the above neutrino mass matrix may be diagonalized approximately with a rough prediction of $m_{\nu_1} \ll m_{\nu_2} \cong m_{\nu_3}$ for the mixing angles of $sin^2\theta_{12} \cong 0.30$, $sin^2\theta_{23} \cong 0.50$, $sin^2\theta_{13} \cong 0.0216$, and $\delta/\pi \cong 1$ or 2, for simplicity, if $-0.4 < \mu/M < 0.3$ and $-0.8 < m/M < -0.4$. We expect to find whether this composite model of neutrino masses and mixing angles is viable, depending on whether the future experimental data on the neutrino masses will agree with this rough prediction.

The original masslessness of neutrinos cannot be explained in the Standard Model, in grand unified theories, or even in supersymmetric grand unified theories, but can be attributed to the degeneracy of the spinor and scalar subquarks of which neutrinos consist in supersymmetric composite models [8]. How to predict and to measure the masses and mixing angles (including the CP violation phase δ) of neutrinos is now one of the hottest subjects in particle physics and is subject to future investigations.

References and Dictionary

References

[1] Nakano T. and Nishijima K., *Prog. Theor. Phys.* **10**, 581(1953); M.Gell-Mann (1953), for a review, see *J. Phys.* **43**, 395(1982).

[2] Gell-Mann M., *Phys. Lett.* **8**, 214(1964); Zweig G. (1964), for a review, see *Int. J. Mod. Phys.* **A25**, 3863(2010).

[3] Nambu Y., in **Preludes in Theoretical Physics**, edited by A.de Shalit (North-Holland, Amsterdam, 1966); Fritzsch H. and Gell-Mann M., in *Proc. XVI International Conf. on High Energy Physics*, edited by Roberts A. (NAL, Batavia, 1973), Vol.2, p.135.

[4] Glashow S. L., *Nucl. Phys.* **22**, 579(1961); Salam A., in *Elementary Particle Physics*, edited by Svartholm N. (Almqvist and Wiksell, Stockholm, 1968), p.367; Weinberg S., *Phys. Rev. Lett.* **19**, 1264(1967).

[5] Cabibbo N., *Phys. Rev. Lett.* **10**, 531(1963); Glashow S. L., Iliopoulos I., and Maiani L., *Phys. Rev. D* **2**, 1285(1968); Kobayashi M. and Maskawa T., *Prog. Theor. Phys.* **49**, 652(1973); Terazawa H., *Prog. Theor. Phys.* **63**, 1779(1980).

[6] Pontecorvo B., *Zh. Eksp. i Teor. Fiz.* **33**, 549(1957)[*Soviet Phys. JETP* **6**, 429(1958)]; **34**, 247(1958)[**7**, 172(1958)]; **53**, 1717(1967)[**26**, 984(1968)]; Maki Z., Nakagawa M., and Sakata S., *Prog. Theor. Phys.* **28**, 870(1962); Terazawa H., *Prog. Theor. Phys.* **57**, 1808(1977); Terazawa H., *Prog. Theor. Phys.* **63**, 1779(1980).

[7] Pati J. C. and Salam A., *Phys. Rev. D* **10**, 275(1974); Terazawa H., Chikashige Y., and Akama K., *Phys. Rev. D* **15**, 480(1977); Terazawa H., *Phys. Rev. D* **22**, 184(1980).

[8] For a classical review, see Terazawa H., in *Proc. 22nd International Conf. on High Energy Physics*, Leipzig, 1984, edited by Meyer A. and Wieczorek E. (Akademie der Wissenschaften der DDR, Zeuthen, 1984), Vol.I, p.63. For more recent reviews, see, for example, Terazawa H., in *Proc. International Conf. "New Trends in High-Energy Physics "*, Alushta, Crimea, 2003, edited by Bogolyubov P. N., Jenkovszky L. L., and Magas V. K. (Bogolyubov Institute for Theoretical Physics, Kiev, 2003), *Ukrainian J. Phys.* **48**, 1292(2003); in *Proc. XXI-st International Conf. "New Trends in High-Energy Physics"*, Yalta, 2007, edited by Bogolyubov P. N., Jenkovszky L. L., and Magas M. K. (Bogolyubov Institute of Theoretical Physics, Kiev, 2007), p.272. For a latest review, see Terazawa H., In *Proc. XXII International Conf. on New Trends in High Energy Physics*, Alushta(Crimea), 2-11, edited by Bogolyubov P. N. and Jenkovszky L. L. (Bogolyubov Institute for Theoretical Physics, Kiev, 2011), p.352, arXiv:1109.3705v5[physics.gen-ph], 10 Feb 2012. For our latest papers on unified subquark models of all fundamental particles and forces, see Terazawa H. and Yasuè M., *J. Mod. Phys.* **5**, 205(2014), arXiv:1401. 3562v8[hep-ph], 5 Apr 2014; *Nonlinear Phenomena in Complex Systems* **19**, 1(2016), arXiv:1505.00172v2[hep-ph] 9 Sep 2015; and Terazawa H., *ibid.* **20**. 327(2017).

[9] Miyazawa H., *Prog. Theor. Phys.* **36**, 1266(1966); Gol'fand Yu. A. and Likhtman L. P., *ZhETF Pis. Red.* **13**, 452(1971)[*JETP Lett.* **13**, 323(1971)]; Volkov D. V. and Aklov V. P., *ibid.* **16**, 621(1972) [*ibid.* **16**, 438(1972); *Phys. Lett.* **46B**, 109(1973); Wess J. and Zumino B., *Nucl. Phys.* **B70**, 39(1974).

[10] Hooft G.'t, in *Recent Developments in Gauge Theories*, edited by Hooft G.'t (Plenum, New York, 1980), Terazawa H., *Prog. Theor. Phys.* **64**, 1763(1980).

[11] Georgi H. and Glashow S. L., *Phys. Rev. Lett.* **32**, 438(1974).

[12] Patrignani C. et al.(Particle Data Group), *Chin. Phys. C* **40**, 100001(2016).

[13] Georgi H., Quinn H. R., and Weinberg S., *Phys. Rev. Lett.* **33**, 451(1974).

[14] Terazawa H., Chikashige T., and Akama K., *Phys. Rev. D* **15**, 480(1977).

[15] Terazawa H., *Prog. Theor. Phys.* **58**, 1276(1977); Visnjic-Triantafillou V., Fermilab Report No. FERMILAB-Pub-80/34-THY, 1980 (unpublished); Terazawa H., *Prog. Theor. Phys.* **64**, 1763(1980); Greenberg O. W. and Sucher J., *Phys. Lett.* **99B**, 339(1981); Terazawa H., *Mod. Phys. Lett.* **A7**, 3373(1992).

[16] Terazawa H., *Phys. Rev. D* **22**, 184(1980).

[17] Terazawa H., *Mod. Phys. Lett.* **A7**, 3373(1992); **A11**, 2463(1996). Recently, a deviation of $\Delta = 0.0083(28)$, which is 3.0 times the stated error, was found from the unitarity condition for the first row of the CKM matrix. This might be taken as an evidence for the substructure of quarks. See Abele H. et al., *Phys. Rev. Lett.* **88**, 211801(2002).

[18] Terazawa H., *Mod. Phys. Lett.* **A7**, 1879(1992).

[19] Terazawa H., *Prog. Theor. Phys.* **64**, 1763(1980).

[20] Buchmuller W., Peccei R. D., and Yanagida T., *Phys. Lett.* **B124**, 67(1983); Barbieri R., Masiero A., and Veneziano G., *ibid.* **B124**, 179 (1983); Greenberg O., Mohapatra R. N., and Yasuè M., *ibid.* **B128**, 65(1983).

[21] Terazawa H., *J. Phys. Soc. Jpn.* **55**, 4249(1986); Terazawa H. and Yasuè M., *Phys. Lett.* **B206**, 669(1988); Terazawa H., in *Perspectives on Particle Physics*, edited by Matsuda S. et al.(World Scientific, Singapore, 1988), p.193.

[22] Terazawa H. and Yasuè M., *Phys. Lett.* **B307**, 383(1993); Terazawa H., *Mod. Phys. Lett.* **A10**, 199(1995).

[23] For a review, see, for example, Gasser J. and Leutwyler H., *Phys. Rep.* **87**, 77(1982).

[24] Nambu Y. and Jona-Lasinio G., *Phys. Rev.* **122**, 345(1961).

[25] Nambu Y., *Nucl. Phys.* **A629**, 3c(1998); **A638**, 35c(1998).

[26] Davidson A., Schwartz T., and Wali K. C., *J. Phys. G* **24**, L55(1998).

[27] Terazawa H., *Mod. Phys. Lett.* **A13**, 2427(1998).

[28] Terazawa H., *Prog. Theor. Phys.* **79**, 738(1988).

[29] Terazawa H., *Phys. Rev. Lett.* **22**, 254(1969); *ibid.* **22**, 442(E)(1969); *Phys. Rev. D* **1**, 2950(1970); Pestieau J. and Roy P., *Phys. Rev. Lett.* **23**, 349(1969); Weinberg S., *Phys. Rev. Lett.* **29**, 388(1972); *Phys. Rev. D* **7**, 2887(1973).

[30] Terazawa H., *Phys. Rev. D* **22**, 2921(1980); **41**, 3541(E)(1990). For the latest discussion, see Terazawa H. and Yasuè M., *J. Mod. Phys.* **5**, 25(2014).

[31] Terazawa H., *Mod. Phys. Lett.* **A5**, 1031(1990).

[32] Kaluza T., *Sitzker. Preuss. Akad. Wiss.* **K1**, 966(1921); Klein O., *Z. Phys.* **37**, 896(1926).

[33] Arkani-Hamed N., Dimopoulos S., and Dvali G., *Phys. Lett.* **B429**, 263(1998); *Phys. Rev. D* **59**, 086004(1999); Antoniadis I. et al., *Phys. Lett.* **B436**, 257(1998); Randall L. and Sundrum R., *Phys. Rev. Lett.* **83**, 3370(1999); **83**, 4690(1999). Note that superstring theories at the Fermi mass scale $G_F^{-1/2}(\sim 300GeV)$ instead of the Planck mass scale $G^{-1/2}(\sim 10^{19}GeV)$ was first suggested by myself by a decade earlier. See Terazawa H., *Prog. Theor. Phys.* **79**, 734(1988).

[34] Terazawa H., Yasuè M., Akama K., and Hayashi M., *Phys. Lett.* **112B**, 387(1982).

[35] Akama K. and Terazawa H., *Phys. Lett.* **B321**, 145(1994), and references therein.

[36] Akama K. and Terazawa H., *Phys. Rev. D* **55**, R2521(1997). Recently, the rare events of 2 electrons + 2 photons + missing transverse energy whose rate was 2.7 sigma above the Standard Model predictions had been reported by the CDF Collaboration. This might be taken as an evidence for the existence of excited leptons or weak bosons. See Loginov A. (for the CDF Collaboration), Report-No. FERMILAB-CONF-05-598-E, hep-ex/0604036, 2006.

[37] Akama K., Katsuura K., and Terazawa H., *Phys. Rev. D* **56**, R2490(1997), and references therein. Recently, the analysis of events with isolated leptons and missing transverse momentum in e^+p collisions by the H1 Collaboration had observed an excess over the Standard Model expectation 3.4

sigma significance. This might be taken as an evidence for the existence of excited leptons or weak bosons. See G.Brandt(for the H1 Collaboration), hep-ex/0701050, 2007.

[38] ATLAS Collaboration, *JHEP* **12**, 55(2015), arXiv:1506.00962v3[hep-ex] 22 Jan 2016; ATLAS Collaboration, ATLAS-CONF-2015-081, Dec. 2015; CMS Collaboration, CMS-PAS-EXO-15-004, Dec. 2015.

[39] Terazawa H. and Yasuè M., *Nonlin. Phenom. Complex Syst.* **19**, 1(2016).

[40] ATLAS Collaboration, *ATLAS-CONF-2017-018*, March 2017; CMS Collaboration, *CMS-PAS-B2G-17-002*, March 2017.

[41] Terazawa H., in *Proc. XXI International Conf. on High Energy Physics*, Paris, 1982, edited by Petiau P. and Porneuf M., *J. de Phys.* **C3**-191(1982).

[42] Terazawa H., in **"What is the electron?"**, edited by Simulik V. (Apeiron, Montreal, 2005), p.29.

[43] For a review, see Terazawa H., in *Proc. International Conf. on Modern Developments in Elementary Particle Physics*, Cairo, Helwan and Assyut, 1999, edited by Sabry A. (Ain Shams University, Cairo, 1999), p.128.

[44] Fukuda Y. et al.(Super-Kamiokande Collaboration), *Phys. Rev. Lett.* **81**, 1562(1998).

[45] Nishijima K., *Phys. Rev.* **108**, 907(1957); Schwinger J., *Ann. Phys.* **2**, 407(1957); Bludman S., *Nuovo Cimento* **9**, 433(1958).

Dictionary

algebra of subquark currents: The algebra consisting of the commutation relations of the currents which are composed of a subquark field and an anti-subquark field.

anomaly-free condition: The condition under which a gauge theory with charged fermions is free from the axial-vector-current anomaly.

CKM matrix: See the **quark-mixing matrix**.

desert hypothesis: The hypothesis that there is no particle with the mass lying between a certain lower energy and another higher one, which was assumed by Georgi, Quinn, and Weinberg in order to explain the disagreement between the experimental value of the weak-mixing angle (θ_w) and the predicted one in the Georgi-Glashow grand unified gauge theory of SU(5)[13].

Gell-Mann matrices: The eight 3×3 matrices (λ's) representing the generators of SU(3), which are extensions of the Pauli matrices of SU(2) or O(3).

Georgi-Glashow relations: The relations between the gauge coupling constants of $SU(3)_c$ strong interactions(f) and $SU(2)_W\times U(1)_Y$ electroweak interactions(g and g') derived by Georgi and Glashow in their unified gauge theory of SU(5) for strong and electroweak interactions of quarks and leptons[11]

lepton-number conservation: The rule that the total lepton number in an initial state is conserved in reactions, which was proposed by Nishijima, Schwinger, and Bludman in 1957 [45]. It holds in strong, electromagnetic, and gravitational interactions, but may violates in weak interactions due to the neutrino oscillation [6], which was found by the Super-Kamiokande Collaboration in 1997 [44].

model of the Nambu-Jona-Lasinio type: The theoretical model in which strong interactions are described by the products of four fermion fields. It was first proposed by Nambu and Jona-Lasinio in their "unified model of elementary particles" in 1961 [24].

Nambu-Goldstone bosons: The scalar and massless bosons which appear due to the spontaneous breakdown of continuous symmetry. Nambu and Jona- Lasinio first demonstrated in their "unified model of elementary particles" of the Nambu-Jona-Lasinio type that the very light pions as composites of nucleon-anti-nucleon pairs can be taken approximately as such Nambu-Goldstone bosons due to the spontaneous breakdown of the chiral symmetry generating the massive nucleons [24].

Nambu-Goldstone fermions: The spinor and massless fermions which appear due to the spontaneous breakdown of supersymmetry.

Nishijima-Gell-Mann rule: The rule of $Q = I_3 + Y/2$ and $Y = B + S$ where Q, I_3, Y, B, and S are the electric charge (in the unit of e), the third component of isospin, "hypercharge", baryon number, and strangeness of a particle, which was proposed by Nakano and Nishijima, and independently by Gell-Mann in 1953 [1].

neutrino mixing: The possibility that the three types of neutrinos (ν_e,ν_μ,ν_τ) in the weak current may not be mass- eigenstates so that they may mix with each other in weak interactions, which was first conjectured by Maki, Nakagawa, and Sakata for the two neutrinos (ν_e,ν_μ) in 1962, and extended to the case of the three or more neutrinos by the author and others in 1980 [6].

neutrino oscillation: The possibility that any one of the three types of neutrinos (ν_e,ν_μ,ν_τ) may change its flavor while traveling due to the neutrino mixing, which was first proposed by Pontecorvo in 1958 [6]

QFD (quantum flavor dynamics): The unified SU(2)$\times$ U(1) gauge theory of elecroweak interactions of quarks and leptons, which was first proposed by Glashow in 1961, and completed with the introduction of the Higgs scalar by Salam and Weinberg in 1967 [4]

quark model for hadrons: The theoretical model of hadrons in which mesons and baryons are taken as color singlet composite states of quark-antiquark pairs and those of three quarks, respectively, which was first proposed by Gell-Mann and Zweig in 1964 [2].

quark-mixing matrix: The $N_g \times N_g$ unitary matrix describing the mixing of the mass eigen-states of the N_g generations of quarks (d,s,b,...) and those in the weak current, which was first introduced by Cabibbo in 1957 and extended by Kobayashi and Maskawa in 1973 [5].

Standard Model: The $SU(3)_c \times SU(2)_W \times U(1)_Y$ gauge theory of strong and electroweak interactions of quarks and leptons, which combines QCD(quantum chromodynamics and QFD (quantum flavor dynamics).

supersymmetry: The hypothetical symmetry between bosons and fermions with the same quantum numbers except for their spins, which was first proposed by Miyazawa in 1966 [9].

weak-mixing angle: The mixing angle for which the third component of $SU(2)_W$ gauge field(A^3_μ) with the coupling constant, g, mixes with the U(1) gauge field(B_μ) with the coupling constant, g', as defined by $tan\theta_w = g'/g$, which was first introduced by Glashow in his unified gauge theory of electroweak interactions in 1961 [4].

Yang-Mills gauge theory: The gauge theory of non-Abelian symmetry, which was first introduced by Yang and Mills for describing strong interactions of baryons with a SU(2) iso-triplet of ρ mesons, and which was later taken not only as a fundamental $SU(2)_W$ gauge theory of weak interactions in QFD(quantum flavor dynamics) by Glashow in 1961 [4] but also as a fundamental $SU(3)_c$ gauge theory in QCD(quantum chromodynamics) by Nambu in 1966 [3].

Picture of Dr. Andrei D. Sakharov (1921-1989) and the Author at the Auditorium, Institute for Nuclear Study, University of Tokyo in 1989.

Interactions with Dr. Andrei D. Sakharov and the Author

1) In 1967, A. D. Sakharov suggested that the Einstein-Hilbert action for gravity can be derived approximately from quantum fluctuations of matter at long distances. A decade later, we demonstrated that not only Einstein theory of gravity in general relativity but also the standard model of strong and electroweak interactions in quantum chromodynamics and in the unified gauge theory can be derived as an effective and approximate theory at low energies from the more fundamental unified composite model of all fundamental particles and forces. Around the same year, he also proposed the three necessary conditions for explaining the baryon number asymmetry in the Universe including the existence of baryon number violating process, the violation of CP invariance in the process, and the expansion of the early Universe (which is out of the thermal equilibrium). Furthermore, he is known to have proposed the muon-catalyzed fusion which may be useful for clean electric generators in the future. No doubt, he was one of the most eminent Russian theoretical physicists in the twentieth century. Unfortunately, he was known as one of the leaders in

making the thermo-nuclear bombs(or Hydrogen bombs) in the Soviet Union. However, It is good for him that he received Nobel Peace Prize in 1975 for his leadership in protecting human-rights in the eastern world during the cold-war era. It would be better for him, however, that he might have received one, two, or even three Nobel Prizes in physics by living much longer!

2) I first met Dr. Andrei Sakharov in December, 1978 when I went to Moscow to present a seminar on the unified theory of all fundamental particles and forces at the First Quantum Gravity Seminar. That was in the midst of "cold time", both literally and politically: the temparature was around 40 degrees below zero in Celsius in Moscow so that some people were freezed to death in the apartments in Moscow because of the shortage of fuel. Also, the people in the western world knew that the famous Russian Nobel Laureate in literature just became in trouble. When Dr. Sakharov appeared at the Seminar cite, he became surounded by many participants from the western world, some of whom dared to ask him whether he had a close contact with the famous novel-writer. It seemed rather fortunate that he had a poor ablity in English conversation while the foreign participants including me had an even poorer one in Russian.

I was honored to have presented an invited talk on the pregeometric theory of gravity in front of the big audience including the originater of pregeometry and, in addition, to have been asked to chair the session for the speakers including such an eminent Russian physicist as Dr. E. S. Fradkin.

It was a pity that Dr. Sakharov himself had not given us any talk in the Seminar probably because he had been too busy to work in physics, being involved in the other fields such as the political movement for protecting human-rights in the Soviet Union. However, we, all the invited participants from the western or far-eastern countries, seemed to be satisfied with the heavy courtesy service which we had received from the organizers of the Seminar sponsored by the Academy of Sciences of U.S.S.R.

3) On January 24, 1980, I suddenly received a telegram from my friends in Lawrence Berkeley Laboratory, which was crying or shouting, "WISH TO SUGGEST CABLES FROM SCIENTIFIC COMMUNITY AND PROFESSIONAL SOCIETIES TO BREZHNEV AND ALEXANDROV PROTESTING TREATMENT OF SAKHAROV. ALSO, USEFUL TO GIVE TEXTS TO NATIONAL MEDIA. IMPORTANT TO DEMONSTRATE INTERNATIONAL RESPONSE. SILENCE COULD HAVE NEGATIVE EFFECT. BEST WISHES, ...". We knew through the world-wide media that Dr. Sakharov, who had been protesting against the invasion of Afghanistan by the U.S.S.R., had

been arrested on the street in Moscow on January 22, and that he had been sent to Gorky City where he was forced to live in an apartment under the control of the KGB. A professor in the Theory Group in the National Laboratory for High Energy Physics who apparently had received a telegram from the same senders with the same message promptly collected eighteen high energy physicists including him and me to sent a telegram to Dr. Alexandrov, the President of the Academy of Sciences of U.S.S.R. with the following message:

"We, under signed, along with many fellow Japanese physicists, hereby express our grave concern over recent reports concerning the status of Dr. Andrei Sakharov, a national of your country. We ask that Dr. Sakharov's honor as an outstanding scientist be preserved and, further, that his basic human rights be fully restored immediately.

We insist that the thoughts and opinions of Dr. Sakharov and any other man can only be judged under the principles of righteousness and freedom. We also point out that your treatment of Dr. Sakharov can only serve to worsen those international tensions which ever more endanger detente between the nations of the world today."

4) In October, 1981, for the first time after the so-called "Sakhrov Problem" had occurred, I went to U.S.S.R. to present a talk with the title of "Pregeometry" at the Second Seminar, "Quantum Gravity", Moscow. October 13-15, 1981. I was invited formally by Prof. M. A. Markov, Chairman, Nuclear Physics Department, Academy of Sciences of U.S.S.R. as in the First Seminar on Quantum Gravity, Moscow, 1978. It was cold in October in Moscow but much warmer than it had been in December, 1978. However, the atmosphere in the Seminar was even much colder than in the First Seminar in 1978 as every participant missed seeing Dr. Sakharov at the Seminar cite. During the Seminar, an identified person approached and said to me, "Are you interested in seeing the wife of Dr. Sakharov at her apartment? I will take you there if you are." I instantly declined his offer by shouting loudly, "No! No! He is one of many Russian friends of mine!", meaning that it might have hurt the other Russian friends including the organizers if I had accepted his offer to escape from the Seminar.

On November 22, 1981, Dr. Sakhalov and his wife, Mrs. Yelena Bonner, began a hunger-strike for demanding their government to issue an exit visa for Miss Liza Alexeyeva, the fiancée of his son-in-law, Mr. Alexey Semyonov, to go to U.S.A. and to live with him in Newton, Massachusetts. They continued the strike for seventeen days, becoming seriously ill and being hospitalized. It was not until on December 9 that the KGB told her that the exit visa would be issued to her. On December 21, she finally arrived in Boston and met her fiancé happily after the three-year's interruption.

5) On May 21, 1982, the 61-st anniversary of Dr. Sakharov's birthday, I was asked to present a talk entitled, "Physics of Dr. Sakharov", in the "Meeting on Sakharov's Birthday" at Tokyo Yamanote Kyoukai, the largest Catholic church in Tokyo. The Meeting was organized by the ten members consisting of the group called "Campaign for Saving Sakharov". Among the six speakers in the Meeting, there were the only two that do not belong to the group, including Dr. Otsuhiko Kaga, MD, the famous novel writer and medical doctor, and me. Also, perhaps, I was the only one speaker that was not a Christian. It was the first and perhaps last time for me to give a speach in a church as if I had been a priest.

In addition to the six talks on the various subjects, ranging from socialism and human-rights to pregeometry and the three conditions for the baryon asymmetry in the Universe, there shown was the very valuable film which had been made and taken out of Gorky City secretly and which showed Dr.Sakharov standing together with his wife and speaking by a Christmas tree in the apartment.

In that morning, this meeting was announced in the greatest details by Mainichi Shinbun, which seems the most liberal among the big three newspapers in Japan.

6) On May 2, 1984, Dr. Sakharov began a hunger-strike again demanding this time an exit visa for his wife, Mrs. Yelena Bonner, to go out of U.S.S.R. in order to have a better medical treatment on her heart. On May 13, she herself also joined the strike. It seemed a matter of time whether either one or both of them would die. That was an enough reason why I had decided on May 19 to sent a telegram to Prof.M.A.Markov, Chairman, Nuclear Physics Department, Academy of Sciences of U.S.S.R. with the following message:

> *"Dear Professor Markov, We, the undersigned, along with many fellow Japanese physicists, hereby express our grave concern over the latest reports concerning the present status of Dr. Andrei Sakharov, a member of your Academy of Sciences. Sincerely yours, ..."*

The undersigned include forty-three Japanese physicists including Prof. M. Koshiba, my life-long tutor, and me.

Fortunately, on Sunday, May 20, The Asahi Shimbun (Newspaper) had a small article on our effort for saving their lives with the title, "Japanese scholars have sent a telegram, too.".

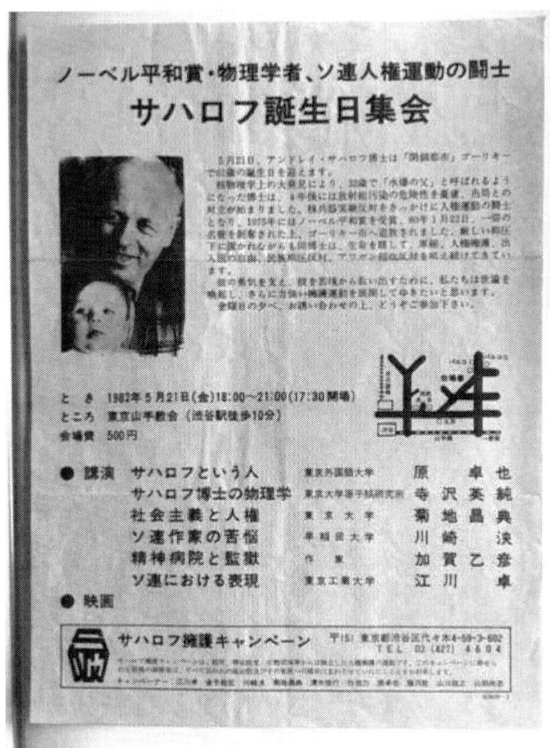

The notice for "Meeting on Sakharov's Birthday", Tokyo Yamanote Kyoukai, May 21, 1982.

7) On June 25, 1984, I was asked to make two press conferences together with Dr. Alexey and Mrs. Liza Semyonov firstly in Japanese at the conference room in the Hall of Members of the House of Councillors (or Upper House of the Diet) of Japan, and secondly in English at the conference room in the Foreign Reporter Club. The purpose for the conference was to let not only all the people in Japan but also all the people in the foreign countries get acquainted with the terrible situation in which Dr. Sakharov and his wife had been forced to live. The night before that day, I went to a certain apartment in Tokyo to see the young Russian-American couple who had just arrived in Tokyo and was handed a copy of the English translation of the latest paper written by A. D. Sakharov with the title of "Cosmological Transitions with a Change in Metric Signature".

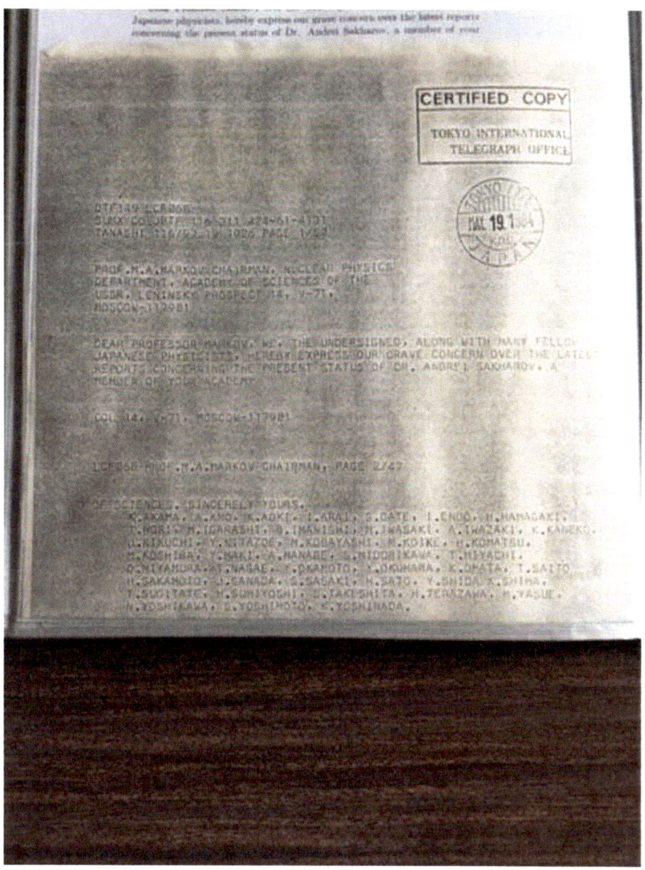

CERTIFIED COPY

TOKYO INTERNATIONAL TELEGRAPH OFFICE

PROF.M.A.MARKOV CHAIRMAN, NUCLEAR PHYSICS DEPARTMENT, ACADEMY OF SCIENCES OF THE USSR, LENINSKY PROSPECT 14, V-71, MOSCOW-117981

DEAR PROFESSOR MARKOV, WE, THE UNDERSIGNED, ALONG WITH MANY FELLOW JAPANESE PHYSICISTS, HEREBY EXPRESS OUR GRAVE CONCERN OVER THE LATEST REPORTS CONCERNING THE PRESENT STATUS OF DR. ANDREI SAKHAROV, A MEMBER OF YOUR ACADEMY

OF SCIENCES. SINCERELY YOURS.

K.AKAMA, A.ANO, K.AOKI, T.ARAI, S.DATE, I.ENDO, H.HAMAGAKI, T.HORI, M.IGARASHI, S.INANISHI, M.IWASAKI, A.IWAZAKI, K.KANEKO, J.KIKUCHI, Y.KITAZOE, M.KOBAYASHI, M.KOIKE, H.KOMATSU, M.KOSHIBA, Y.MAKI, A.MANABE, S.MIDORIKAWA, T.MIYACHI, O.MIYAMURA, T.NAGAE, T.OKAMOTO, Y.OKUHARA, K.OHATA, T.SAITO, H.SAKAMOTO, J.SANADA, S.SASAKI, H.SATO, Y.SHIDA, K.SHIMA, T.SUGITATE, H.SUMIYOSHI, S.TAKESHITA, H.TERAZAWA, M.YASUE, K.YOSHIKAWA, I.YOSHIMOTO, K.YOSHINADA.

The telegram sent from the 43 Japanese physicists to Prof.M.A.Markov, Chairman, Nuclear Physics Department, Academy of Sciences of U.S.S.R. on May 19, 1984.

I was asked to read the paper and to announce at the press conferences on the next day that Dr. Sakharov seemed to be mentally normal and vivid in spite of the most isolated situation in which he was forced to work. After reading it at my home, I found that the paper clearly showed the same eminent originality as we had found in the legendary papers published in 1967!

Both of the press conferences turned out to be very successful. We had many reportors not only from newspaper companies but also from TV broadcasting companies. After I came back home, I had a strange feeling in watching my face zoomed up on TV in the most popular news program of Nippon(or Yomiuri) TV Broadcasting during the golden hour of the day.

Picture of Dr. Alexey and Mrs. Liza Semyonov together with the Author, his wife, and daughter, on June 24, 1984.

8) In July, 1984, I went to Leipzig for the XXII International Conference on High Energy Physics, Leipzig, July 19-25, 1984. I was invited by the East German organizers to organize the parallel session of "Sub-component Models". Before the Conference, all the invited organizers of the parallel sessions and the invited rapporteurs in the plenary sessons were requested to attend the pre-conference in order to prepare for those sessions. As I was walking on the street in the lovely city on the day of my arrival, I met Professor Masatoshi Koshiba, my life-long tutor since my undergraduate studies at the University of Tokyo, who was invited to be the rapporteur in the plenary session of "Proton Decay Experiments". We were only two physicists coming from the far-east country to the East Germany for the pre-conference.

The Conference was very busy and exciting since it was the most official conference in high energy physics to be held in every other year and since it was the first one just after the great discovery of the weak bosons which had been predicted in the unified gauge theory of electroweak interactions for the last more than two decades.

During the Conference, one of my American friends, who had sent a telegram to me from the Lawrence Berkeley Laboratory in 1980, was making every effort to collect many signatures from the participants for the statement to protest against the Soviet Union for saving the life of Dr. Sakharov. I signed but told him not to make the statement in public until all the participants who had signed had gone out of the eastern countries under the control of the Soviet Union as, otherwise, they would be in danger. He agreed with me and kept his promise to make it in public in the western countries after the Conference.

9) On April 16, 1985, Dr. Sakharov began a hunger-strike once again but was forced to have meals after he was hospitalized on April 21. It was reported that he would quit a member of the Academy of Sciences of U.S.S.R. on April 10 unless the Academy should make an effort for improving the situation from which he and his wife had been suffering. It was well known that the Academy was the only one domestic organization that was supporting him not only financially but also spiritually.

On May 18, 1985, together with my fifteen colleagues at Institute for Nuclear Study, University of Tokyo, I sent a telegram again to Prof. M. A. Markov, Chairman, Nuclear Physics Department, Academy of Sciences of the U.S.S.R. with the following message:

> *"Dear Prof. Markov, We, the under signed, along with many fellow Japanese physicists, hereby express our grave concern over the latest reports concerning the request by Dr. Sakharov to your Academy. We, nuclear physicists, ever more highly respect the conscience of the eminent Russian colleague, for all lives in the world may depend also on our conscience in this tough nuclear era. Sincerely yours, ..."*

On December 16, 1986, Dr. Sakharov was suddenly called up by Mr. Gorbachov from Kremlin and was told to be released from Gorky City. On December 23, he finally returned to his apartment in Moscow after the long exile to Gorky City for almost seven years. Also, his wife became well after having received the best medical treatments out of the Soviet Union. The millions of cordial prays for them and the thousands of warm aids to them from all over the

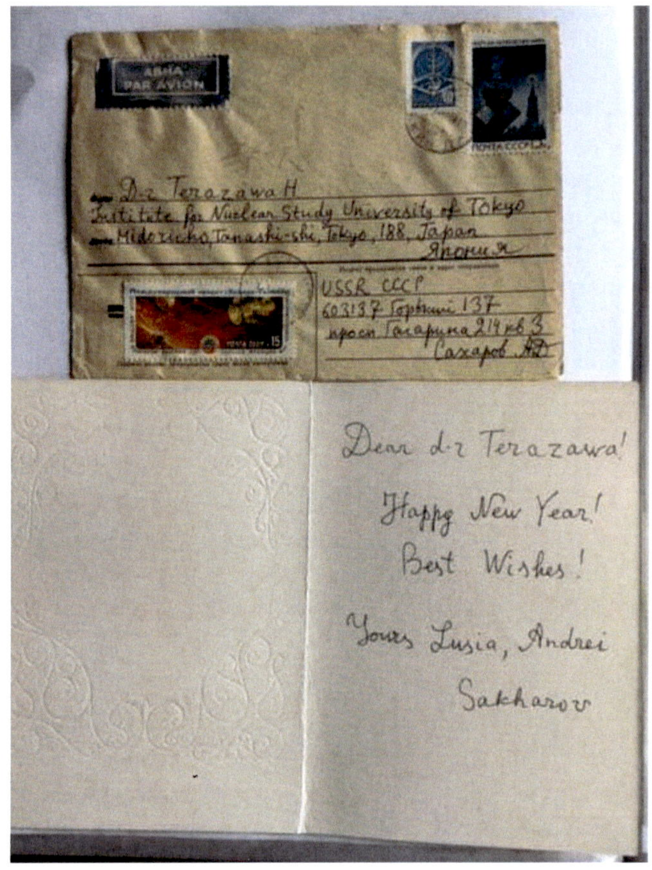

The Christmas card from Dr. Sakharov to the Author in December, 1984.

world have worked!

10) On December 29, 1986, The Tokyo Shimbun (newspaper) presented the top article on the first page with the large title of "Interview of Dr. Sakharov". In the interview held in his apartment in Moscow on December 25, he talked about their hard and long exile life of total isolation in Gorky City, his opinion on the Afghanistan problem, his intention of going out of the Soviet Union, etc. He also added that he had been delighted to receive a new-year's card from Prof. Hidezumi Terazawa and that he could survive thanks to the help by his wife.

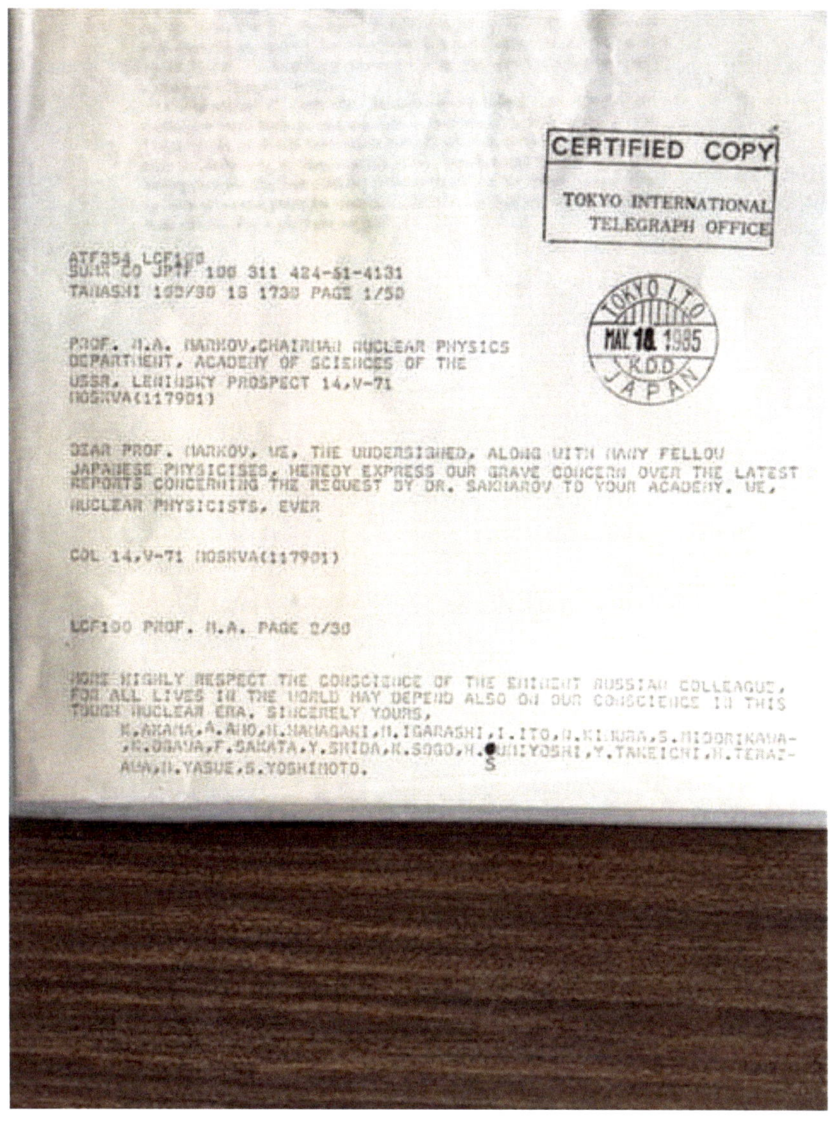

CERTIFIED COPY
TOKYO INTERNATIONAL TELEGRAPH OFFICE

TOKYO I.T.O.
MAY 18 1985
K.D.D.
JAPAN

ATF354 LCF100
SUMR CO JPTF 100 311 424-51-4131
TANASHI 100/30 18 1730 PAGE 1/50

PROF. M.A. MARKOV,CHAIRMAN NUCLEAR PHYSICS
DEPARTMENT, ACADEMY OF SCIENCES OF THE
USSR, LENINSKY PROSPECT 14,V-71
MOSKVA(117901)

DEAR PROF. MARKOV, WE, THE UNDERSIGNED, ALONG WITH MANY FELLOW
JAPANESE PHYSICISES, HEREBY EXPRESS OUR GRAVE CONCERN OVER THE LATEST
REPORTS CONCERNING THE REQUEST BY DR. SAKHAROV TO YOUR ACADEMY. WE,
NUCLEAR PHYSICISTS, EVER

COL 14,V-71 MOSKVA(117901)

LCF100 PROF. M.A. PAGE 2/30

MORE HIGHLY RESPECT THE CONSCIENCE OF THE EMINENT RUSSIAN COLLEAGUE,
FOR ALL LIVES IN THE WORLD MAY DEPEND ALSO ON OUR CONSCIENCE IN THIS
TOUGH NUCLEAR ERA. SINCERELY YOURS,
K.AKAMA,A.ANO,M.NAKAGAKI,M.IGARASHI,I.ITO,M.KIMURA,S.MIDORIKAWA-
,K.OGAWA,F.SAKATA,Y.SHIDA,K.SOGO,H.SUMIYOSHI,Y.TAKEICHI,H.TERAZ-
AWA,M.YASUE,S.YOSHIMOTO.

The telegram sent from the 16 Japanese physicists to Prof.M.A.Markov, Chairman, Nuclear Physics Department, Academy of Sciences of U.S.S.R. on May 18, 1985.

According to the Issue Vol. 3, Nos. 4, 5 (April/May 1985) of SAMIZDAT: Inner Report from USSR, the popular newspaper in West Germany announced that they had obtained two films showing Dr. Sakharov and that the one of these films shows him having a letter from Prof. Hidezumi Terazawa, Institute of Nuclear Study, University of Tokyo, The Report added that, if it was true, there seemed to be a way for letters or messages to reach him in Gorky City. On October 22, 1988, The Asahi Shimbun put a big article on the first page with the title of "Dr. Sakharov Going to U.S.A. Next Month". It reported that he was allowed to go out of the Soviet Union for his first time in his whole life in order to attend an international conference on human life and development in November in U.S.A.!

On August 18, 1989, The Yomiuri Shimbun, the most conservative among the big three newspapers in Japan, put an article in the center of the first page announcing the Second Nobel Laureate Japan Forum, Tokyo and Sapporo, 1989 with the title, "Dr. Sakharov, attending with his wife, Yelena Bonner, coming to Japan this fall for the first time". It also spent two additional pages for many articles on their coming first visit to Japan. According to the one of those articles, he had three desires to do during their coming stay in Japan: to see Mt. Fuji, the highest and most famous mountain in Japan; to visit Hiroshima by Shinkansen, the fastest and most famous train in the world; and to meet Prof. Hidezumi Terazawa, his longest Japanese friend!

11) On November 27, 1989, there appeared the opening ceremony of the Second Nobel Laureate Japan Forum, 1989 at the largest Banquet Hall of the Imperial Hotel. I was invited by the organizers of the Forum sponsored mainly by The Yomiuri Shimbun to the Banquet. Right after I arrived at the Banquet cite, I was met by the head of science column of the newspaper and taken to see Dr. Sakharov for the second time after the long interruption of almost eleven years! We shook hands and said, "Nice to see you again" in the flashlights of newspaper cameras and in the dazzling lights of TV cameras.

On November 6, 1989, Dr. Sakharov visited Institute for Nuclear Study, University of Tokyo, and gave us a colloquium on "Perestroika", according to my old invitation. In the early morning of the day, I went to the Institute and found the big car parking in front of the Main Entrance, which meant that Dr. Sakharov had already been there. Then, I saw him coming back from a morning work, being associated with the able secretary of the Director of the Institute. It was the first time for the Auditorium of the Institute to be over-crowded by hundreds of physicists, engineers, technical staffs, secretaries, and others who wanted to listen to his speech in Russian, the totally foreign language to Japanese.

Later, I was asked by Koudansha, one of the largest publishing company in Japan, to interview Dr. Sakharov in the Imperial Hotel where he and his wife were staying. My wife, Keiko, was also invited by his wife, but she declined it as she had a cold. I met Dr. Sakharov, his wife, the editor and writer of Koudansha, and the famous Russian-Japanese simultaneous translator at a cozy hotel room. Dr. Sakharov talked in Russian on physics and politics almost all by himself for an hour and the translator did her job perfectly well so that they might not need my help at all. Later, I received a copy of certain issue of the popular scientific journal published monthly by Kodansha, which reported the interview of Dr. Sakharov by me and showed the lovely pictures of Dr. Sakharov, his wife, and me.

Less than a month had passed since he had been back to Moscow from Tokyo. On December 14, he suddenly died (probably due to the heart attack) on the bed in his apartment in Moscow, at his age of sixty-eight.

12) In 1991, I went to Moscow for the First A. D. Sakharov International Conference on Physics, Moscow, May 27-31, 1991, organized by the Academy of Sciences of USSR. I was invited to present a talk with the title of "Future Prospects of Pregeometry" in the plenary session. It was wonderful that Prof. K. Nishijima, a foreign member of the Academy and one of the six tutors in my graduate study in University of Tokyo, was invited not only to give a talk on quark confinement in QCD, but also to chair the plenary session of Prof. Y. Ne'eman and his old student, me! Also, Prof. K. Akama, one of my life-long collaborators, were invited to give a talk on his new theory of pregeometry.

During the Conference, I was asked by Ms. Yelena Bonner to fly to the south from Moscow to a certain state in U.S.S.R. for watching what was going on there under the power of the Soviet Union. Fortunately or unfortunately, I could not fly out of Moscow Region because of my restrictive visa. However, I agreed with her at the statement for the people in the state and against the Soviet Union which were approved unanimously in the closing plenary session on the last day of the Conference. All the participants saw her very vivid and still working very hard for human rights, following the will of her late husband.

The Conference was dedicated to the memory of Dr. Sakharov, the great Russian not only physicist but also human-right saver so that hundreds of physicists in all the fields got together, which was rather unusual in those days. The contents of invited talks had been published in the two volumes of "Sakharov

Memorial Lectures in Physics", the Proceedings of the First International A. D. Sakharov Conference on Physics, Moscow, 1991, edited by L. V. Keldysh and V. Ya. Feinberg (Nova Science, New York, 1992)!

I think it needless to tell you, the readers, about what happened in the world, right after that.

Chapter III

Dark Energy, Dark Matter, and Strange Stars*

Abstract

A theory of special inconstancy, in which some fundamental physical constants such as the fine-structure and gravitational constants may vary, is proposed in pregeometry. In the special theory of inconstancy, the $\alpha - G$ relation of $\alpha = 3\pi/[16\ln(4\pi/5GM_W^2)]$ between the varying fine-structure and gravitational constants (where M_W is the charged weak boson mass) is derived from the hypothesis that both of these constants are related to the same fundamental length scale in nature. Furthermore, it leads to the prediction of $\dot{\alpha}/\alpha = (-0.8 \pm 2.5) \times 10^{-14} yr^{-1}$ from the most precise limit of $\dot{G}/G = (-0.6 \pm 2.0) \times 10^{-12} yr^{-1}$ by Thorsett, which is not only consistent with the recent observation of $\dot{\alpha}/\alpha = (0.5 \pm 0.5) \times 10^{-14} yr^{-1}$ by Webb et al. but also feasible for future experimental tests. In special inconstancy, the past and present of the Universe are explained and the future of it is predicted, which is quite different from that in the Einstein theory of gravitation. Also, a theory of general inconstancy, in which any fundamental physical constants may vary, is proposed

---

*Some contents of this Chapter have been presented in the key note entitled, "Varying Fundamental Physical Constants – Dark Energy, Dark Matter, and Strange Stars(*Dedicated to Professor Hironari Miyazawa*)", and contributed to the 9th Bolyai-Gauss-Lobachevsky International Conference on Non-Euclidean Geometry in Modern Physics and Mathematics, Minsk, Belarus, 2015, published in the Proceedings, edited by Yu. Kurochkin and V. Red'kov (Institute of Physics, National Academy of Sciences of Belarus, Minsk, 2016), p.295.

in "more general relativity", by assuming that the space-time is "environment-dependent". In the general theory of inconstancy, the $G-\Lambda$ relation between the varying gravitational and cosmological constants is derived from the hypothesis that the space-time metric is a function of τ, the "environment coordinate", in addition to x^{μ}, the ordinary space-time coordinates. Furthermore, it leads to the prediction of the varying cosmological constant, which is consistent with the present observations. Finally, simple solutions to the most intriguing problems in current cosmology, the problems of the dark energy and dark matter, are proposed in the theory of general inconstancy and in the so-called Bodmer-Terazawa-Witten hypothesis of strange matter.

This Chapter consists of the following Sections: I. Introduction, II. Pregeometry, III. Special Inconstancy, IV. General Inconstancy, and V. Dark Energy, Dark Matter, and Strange Stars – Future Prospects.

It also contains References and a Dictionary for Chapter III.

I. Introduction

Is a physical constant really constant? In 1937, Dirac [1] discussed possible time variation in the fundamental constants of nature. He made not only the **large number hypothesis(LNH)** but also, as a consequences of the LNH, the astonishing prediction that the gravitational constant G varies as a function of time. Since then, Jordan [2] and many others [3,4] have tried to construct new theories of gravitation or general relativity in order to accommodate such a time-varying G. Although the LNH has been inspiring many theoretical developments and has recently led myself [5] to many **new large number relations**, the prediction of the varying G has not yet received any experimental evidence. Recently, Thorsett [6] has shown that measurements of the masses of young and old neutron stars in pulsar binaries lead to the most precise limit of

$$\dot{G}/G = (-0.6 \pm 2.0) \times 10^{-12} yr^{-1}$$

at the 68% confidence level.

More recently, on the other hand, Webb et al. [7] have investigated possible time variation in the fine structure constant α by using quasar spectra over a wide range of epochs, spanning redshifts $0.2 < z < 3.7$, in the history of our Universe, and derived the remarkable result of

$$\dot{\alpha}/\alpha = (6.40 \pm 1.35) \times 10^{-16} yr^{-1}$$

for $0.2 < z < 3.7$, which is consistent with a time-varying α. Note, however, that in 1976 Shylakhter [8] obtained the very restrictive limit of $|\Delta\alpha/\alpha| < 10^{-7}$ or, more precisely,

$$\Delta\alpha/\alpha\Delta t = (-0.2 \pm 0.8) \times 10^{-17} yr^{-1}$$

for $z \sim 0.16$ (but over a narrower and latest range of epochs between now and about 1.8 billion years ago) from the "Oklo natural reactor". Very lately, Srianand et al. [9] have made a detailed many-multiplet analysis performed on a new sample of Mg II systems observed in high quality quasar spectra obtained using the Very Large Telescope and found a null result of $\Delta\alpha/\alpha = (-0.06 \pm 0.06) \times 10^{-5}$ for the fractional change in α or 3σ constraint of

$$-2.5 \times 10^{-16} yr^{-1} \leq (\Delta\alpha/\alpha\Delta t) \leq +1.3 \times 10^{-16} yr^{-1}$$

for $0.4 \leq z \leq 2.3$, which seems to be inconsistent with the result of Webb et al. [7]. However, a careful comparison of these different results [7-8] indicates that they are all consistent with a time-varying α as

$$\dot{\alpha}/\alpha = (0.5 \pm 0.5) \times 10^{-14} yr^{-1}$$

for $2.2 < z < 3.7$. More lately, Kanekar et al. have constrained the fundamental constant evolution with HI and OH lines as $\Delta\alpha/\alpha = (-1.7 \pm 1.4) \times 10^{-6}$ over a look-back time of 6.7Gyrs or, equivalently,

$$\dot{\alpha}/\alpha = (-2.5 \pm 2.1) \times 10^{-16} yr^{-1}$$

for $0 \leq z \leq 0.765$ [8].

In this Chapter, I am going to introduce a theory of special inconstancy, in which some fundamental physical constants such as the fine-structure and gravitational constants may vary. In the special theory of inconstancy, the $\alpha - G$ relation of

$$\alpha = 3\pi/[16\ln(4\pi/5GM_W^2)]$$

(where M_W is the charged weak boson mass) is derived from the hypothesis that both of α and G are related to the same fundamental length scale in nature. Furthermore, from the limit on $\dot{G}$ by Thorsett, it leads to the prediction of

$$\dot{\alpha}/\alpha = (-0.8 \pm 2.5) \times 10^{-14} yr^{-1}$$

which is not only consistent with the above result on $\dot{\alpha}$ but also feasible for future experimental tests. In special inconstancy, the past and present of the Universe are explained and the future of it is predicted, which is quite different from that in the Einstein theory of gravitation. Also, a general theory of inconstancy, in which any fundamental physical constants may vary, is proposed in "more general relativity", by assuming that the space-time is "environment- dependent". In the general theory of inconstancy, the "$G-\Lambda$ relation" between the varying gravitational and cosmological constants is derived from the hypothesis that the space-time metric is a function of τ, the "environment coordinate", in addition to x^{μ}, the ordinary space-time coordinates. Furthermore, it leads to the prediction of the varying cosmological constant, which is consistent with the present experimental observations.

I will organize this Chapter as follows: in Section II, I will briefly review pregeometry in which a theory of special inconstancy is constructed. In Sections III and IV, I will present the theories of special and general inconstancy and their predictions, respectively. Finally in Section V, I will discuss future prospects by proposing simple solutions to the most intriguing problems in current cosmology, the problems of dark energy and dark matter, in my theory of general inconstancy and in the so-called Bodmer-Terazawa-Witten hypothesis of strange quark matter.

II. Pregeometry

Pregeometry is a theory in which Einstein geometrical theory of gravity in general relativity can be derived from a more fundamental principle as an effective and approximate theory at low energies (or at long distances). In 1967, Sakharov [10] suggested possible approximate derivation of the Einstein-Hilbert action from quantum fluctuations of matter. A decade later, we [11] demonstrated that not only Einstein theory of gravity in general relativity but also the standard model of strong and electroweak interactions in quantum chromodynamics and in the unified gauge theory can be derived as an effective and approximate theory at low energies from the more fundamental unified composite model of all fundamental particles and forces [12].

Let us explain what pregeometry means more explicitly in a simple model of

$$S_0 = \int d^4x\sqrt{-g}L_0(g_{\mu\nu}(x), A_\mu(x), \phi_i(x))$$

where $g_{\mu\nu}$ is the space-time metric, $g = det(g_{\mu\nu})$, A_μ is an Abelian gauge field, and ϕ_i $(i = 1 \sim n)$ are n complex scalar fields of matter with the charge e. The fundamental Lagrangian L_0 consists of the gauge-invariant kinetic terms of the matter fields only as

$$L_0 = g^{\mu\nu}[(\partial_\mu + iA_\mu)\phi_i^\dagger](\partial_\nu - iA_\nu)\phi_i - F^{-1}$$

(where F is an arbitrary constant) but does not contain either the kinetic term of the space-time metric or that of the gauge field so that both of $g_{\mu\nu}$ and A_μ are auxiliary fields. The effective action for the space-time metric and gauge field can be defined by the path-integral over the matter fields as

$$exp(iS_{eff}) = \int \prod_i [d\phi_i^\dagger][d\phi_i]exp(iS_0)$$

and it can be expressed formally as

$$S_{eff} = -iTrln[(\partial_\nu - iA_\nu)\sqrt{-g}g^{\mu\nu}(\partial_\mu + iA_\mu)]$$
$$-\int d^4x\sqrt{-g}F^{-1}$$

after the path-integration over ϕ_i. For small scalar curvature R and Ricci curvature tensor $R_{\mu\nu}$, the effective action can be calculated to be

$$S_{eff} = \int d^4x\sqrt{-g}[2\lambda + (1/16\pi G)R + c(R^2 + dR^{\mu\nu}R_{\mu\nu})$$
$$+(1/4e^2)F^{\mu\nu}F_{\mu\nu} + ...]$$

with

$$2\lambda = [n\Lambda^4/8(4\pi)^2] - F^{-1},$$
$$(1/16\pi G) = n\Lambda^2/24(4\pi)^2,$$
$$c = nln\Lambda^2/240(4\pi)^2,$$
$$d = 2,$$

and

$$(1/4e^2) = nln\Lambda^2/3(4\pi)^2,$$

where λ and Λ are the cosmological constant and the **momentum cut-off of the Pauli-Villars type**, respectively. Notice that the arbitrary constant F^{-1} plays a role of counter term so that the cosmological constant may become as small as or as large as it is observed. Notice also that the momentum cut-off Λ must be of order of the Planck mass $G^{-1/2}(\sim 10^{19}GeV)$. Furthermore, not only the R^2 and $R^{\mu\nu}R_{\mu\nu}$ terms but also the remaining terms in the expansion of S_{eff} are practically negligible. This complete a simple demonstration that not only the **Einstein-Hilbert action of gravity** but also the **Maxwell action of electromagnetism** in general relativity can be derived as an effective and approximate theory at low energies from the simple model in pregeometry, provided that there exists a natural momentum cut-off at around the Planck mass in nature [13].

One of the most remarkable consequences of pregeometry is the $\alpha - G$ relation, a simple relation between the fine-structure and gravitational constant, which can be easily derived from the results for α and for G by eliminating the momentum cut-off Λ. In our unified quark- lepton model of all fundamental forces [14,15], the $\alpha - G$ relation is given by [16]

$$\alpha = 3\pi / \sum_i Q_i^2 ln(12\pi/nGm_i^2),$$

where Q_i and m_i are the charge and mass of quarks and leptons, respectively. For three generations of quarks and leptons and their mirror- or super-partners, the $\alpha - G$ relation simply becomes

$$\alpha \cong 3\pi/16ln(4\pi/5GM_W^2)$$

where M_W is the charged weak boson mass. Notice that this $\alpha - G$ relation is very well satisfied by the experimental data of $\alpha \cong 1/137$, $G^{-1/2} \cong 1.22 \times 10^{19}GeV$, and $M_W \cong 80.4GeV$ [17].

III. Special Inconstancy

Special inconstancy is a principle in which some fundamental physical constants such as the fine-structure and gravitational constants may vary. Let us first make it clear that in this note we use the natural unit system of $h/2\pi = c = 1$ (where h is the Planck constant and c is the speed of light in vacuum). Notice, however, that it does not mean that in discussing the relevant possibility of the varying fine-structure and gravitational constants [18], we exclude another intriguing possibility of the varying light velocity recently discussed by some authors [19] since varying either h or c is inevitably related to varying the fine-structure constant $\alpha(\equiv e^2/2hc)$(if the unit charge e stays constant). It simply means that we must set up a certain reference frame on which we can discuss whether physical quantities such as the fine-structure, gravitational, and cosmological [20] constants be really constant. Our basic hypothesis is that both of the fine-structure and gravitational constants are related to the more fundamental length scale of nature as in the unified (pregauge [21] and) pregeometric [10-12] theory (or "pregaugeometry" in short) of all fundamental forces [14, 15] reviewed in the last Section.

To be more explicit, in the simple model of pregaugeometry discussed in the last Section, assert that

$$< (\partial_\mu + iA_\mu)\sqrt{-g}g^{\mu\nu}(\partial_\nu - iA_\nu)\phi_i >_\Lambda = 0,$$

$$g_{\mu\nu} = F < [(\partial_\mu + iA_\mu)\phi_i^\dagger](\partial_\nu - iA_\nu)\phi_i >_\Lambda,$$

and

$$A_\mu = (i/2) < [\phi_i^\dagger \partial_\mu \phi_i - (\partial_\mu \phi_i^\dagger)\phi_i]/(\phi_j^\dagger \phi_j) >_\Lambda,$$

where $<>_\Lambda$ denotes the expectation value in the space-time with the fundamental length scale parameter of Λ^{-1}. The first equation is the usual field equation for ϕ_i while the last two can be taken either as the "equations of motion" for $g_{\mu\nu}$ and A_μ, which can be derived from the fundamental action S_0, or as the "fundamental field equations", which can reproduce the effective Einstein-Hilbert-Maxwell action S_{eff} at low energies($\ll \Lambda$) or at long distances($\gg \Lambda^{-1}$).

The most important consequence of special inconstancy in pregaugeometry is the $\dot{\alpha} - \dot{G}$ relation for the varying fine-structure and gravitational constant of

$$\dot{\alpha}/\alpha^2 = (16/3\pi)[(\dot{G}/G) + 2(\dot{M}_W/M_W)],$$

which can be derived from differentiating both hand sides of the $\alpha - G$ relation with respect to any parameter for varying fundamental physical constants. This immediately leads to the remarkable predictions of

$$\dot{\alpha}/\alpha = (-0.8 \pm 2.5) \times 10^{-14} yr^{-1}$$

for constant M_W and

$$\dot{M}_W/M_W = (0.5 \pm 1.2) \times 10^{-12} yr^{-1}$$

from the limit of $\dot{G}/G = (-0.6 \pm 2.0) \times 10^{-12} yr^{-1}$ by Thorsett [6] and from the experimental data of $\dot{\alpha}/\alpha = (0.5 \pm 0.5) \times 10^{-14} yr^{-1}$ by Webb et al. [7]. The first prediction is not only consistent with the experimental data by Webb et al. [7] but also feasible for future experimental tests. The second prediction, however, seems too small to be feasible for experimental tests in the near future although such prediction for the possible varying particle masses seems extremely interesting at least theoretically. Notice that the varying M_W is perfectly possible through the varying electroweak gauge coupling constant g (which is related to the fine-structure constant in the standard unified electroweak gauge theory of Glashow-Salaam-Weinberg [22]) and/or the varying vacuum expectation value of the Higgs scalar v (which is related to the momentum cut-off Λ in the unified composite model of the Nambu-Jona-Lasinio type for all fundamental forces [23]) since $M_W = gv/2$. We must mention, however, that neither one of these predictions may assert for the fine-structure constant α or the weak boson mass M_W to vary since both of these predicted values are consistent with zero within their errors.

Let us now first add that in some pregaugeometric model [24], the $\alpha - G$ relation is not of the type of $\alpha \sim 1/ln(1/GM^2)$ but of the type of $\alpha \sim GM^2$(where M is a parameter of mass dimension) so that the $\dot{\alpha}$-$\dot{G}$ relation becomes

$$\dot{\alpha}/\alpha = (\dot{G}/G) + 2(\dot{M}/M).$$

This type of relation predicts

$$\dot{\alpha}/\alpha = (-0.6 \pm 2.0) \times 10^{-12} yr^{-1}$$

for constant M and

$$\dot{M}/M = (0.3 \pm 1.0) \times 10^{-12} yr^{-1}$$

from the limit by Thorsett [6] and from the experimental data by Webb et al. [7]. We suspect, however, that the predicted value for $\dot{M}/M$ seems too small to be feasible for experimental tests in the near future although the one for $\dot{\alpha}/\alpha$ is consistent with the experimental data by Webb et al. [7]. We must also mention that both of these predicted values are consistent with zero within their errors so that neither the fine-structure constant α nor the mass parameter M may be predicted to vary.

Next, remember that in the principle of special inconstancy we do not assert that physical constants may vary as a function of time but do that they may vary in general, depending on any parameters including the cosmological time, temperature, etc. What is the origin of varying the physical constants? The answer to this question may be related to the answer to another fundamental question: What is the origin of the fundamental length scale Λ^{-1} in nature? It can be spontaneous breakdown of scale-invariance in the Universe, which has been proposed by myself [5] for the last quarter century. It can be the natural, dynamical, automatic, *a priori*, but somewhat "wishful-thinking" cut-off at around the Planck length $G^{1/2}$ where gravity would become as strong as electromagnetism, which was suggested by Landau [13] in 1955. It can also be due to the **Kaluza-Klein extra dimension** [25], which is supposed to be compactified at an extremely small length scale of the order of $G^{-1/2}$ or at a relatively large length scale of the order of 1/TeV recently emphasized by Arkani-Hamed et al. [26]. It seems, however, the most natural and likely that the origin of the fundamental length comes from the substructure of fundamental particles including quarks, leptons, gauge bosons, Higgs scalars, etc. [27, 28]. In the unified composite model of all fundamental particles and forces [28], the fundamental energy scale Λ in pregaugeometry can be related to some even more fundamental parameters such as the masses of subquarks, the more fundamental constituents of quarks and leptons, and the energy scale in quantum subchromodynamics, the more fundamental dynamics confining subquarks into a quark or a lepton. In either way, the fundamental length scale Λ^{-1} can be identified with the size of quarks and leptons, the fundamental particles.

In pregeometric special inconstancy, let us now briefly explain the past and present of the Universe and try to predict the future of it, which may differ from that in the conventional Einstein theory of gravitation in general relativity [29]. The history of our Universe goes as follows: Long, long time ago there was no physical space-time, in which the space-time metric was finite and non-vanishing so that the distance was well defined, but the only matter "existed" in the mathematical space-time. Suddenly, there appeared the big bang of our Universe as a phase transition of the space-time from the **pregeometric phase** to the geometric one due to quantum fluctuations of matter, as suggested by us [30] in the early nineteen eighties, and our Universe had happened to be either flat or open. Then, not only all fundamental particles but also all fundamental forces between them were created and they started obeying the effective theory of all fundamental particles and forces including the Einstein theory of gravity with the non-vanishing and varying cosmological constant. In the earliest era during which the matter density had been extremely small, our Universe had been expanding almost exponentially. It had been the "almost inflationary Universe". In the next era of the radiation dominated Universe, our Universe was expanding less fast. Furthermore, in the last era of the matter dominated Universe, our Universe has still been expanding even faster. This history of our Universe is well simulated by a simple model of $(\Omega_m, \Omega_\lambda, -q) = (0,1,1)$, $(1/4, 3/4, 1/3)$, or $(1/4, 3/4, 1/2)$ for the early inflationary era, for the radiation dominated era, or for the matter dominated era, respectively, where Ω_m, Ω_λ, and q are the "pressureless-matter-density", "scaled cosmological constant", and deceleration parameter of the Universe, respectively. Notice that there must be another "phase transition" in which Ω_λ changed from 1 to 3/4 in between the early inflationary era and the radiation dominated era. Concerning the cosmological constant, I have been most impressed by the recent observation of the "farthest supernova ever seen" by Hubble Space Telescope [31]. "This supernova shows us the universe is behaving like a driver who slows down approaching a red stop light and then hits the accelerator when the light turns green." Notice that this behavior of the Universe is what our model simulates. Note also that our model of the Universe is consistent with the recent measurement of the cosmological mass density from clustering in the Two-Degree-Field Galaxy Redshift Survey [32] which strongly favors a low density Universe with $\Omega_m \cong 0.3$. Very recently, the CBI Collaboration [33,17] has found $\Omega_\lambda = 0.64 + 0.11/-0.14$ and $\Omega_m + \Omega_\lambda = 1.006 \pm 0.006$. More recently, the Wilkinson Microwave Anisotropy Probe (WMAP) team [34,17] has found 1) the first generation of stars to shine in the Universe first ignited only 200 million years after the big bang, 2) the age of the Universe(t_0) is 13.69 ± 0.13 billion years old, and 3) $\Omega_m = 0.26 \pm 0.02$ and $\Omega_\lambda = 0.74 \pm 0.03$.

The future of the Universe in our special inconstant picture can be quite different from that in the Einstein-Friedmann picture: 1) Since the cosmological constant may vary in special inconstancy, the space-time of our Universe which is almost flat and expanding faster and faster may not continue to be flat and accelerating forever. Our Universe may even encounter a "topological phase transition", which was discussed by Wheeler [35] in 1959, from the open Universe to the closed one. 2) If the gravitational constant increases, the expansion of the space-time may not continue forever. The Universe may well stop expanding, start contracting, and even be bouncing forever. If G decreases, it will be more accelerated ever. 3) If the fine-structure constant (and/or other fundamental coupling constants such as the strong and weak coupling constants) varies, our Universe may encounter as "obsolete phase transition" from the matter-dominated Universe to the radiation- dominated one. In short, we can expect anything about the future of our Universe or, in other words, we can predict nothing definite on the destiny of our Universe.

IV. General Inconstancy

General inconstancy is a principle in which any fundamental physical constants may vary. In order to accommodate general inconstancy with general relativity, let us assume that the space-time is environment dependent. More explicitly, let us make the hypothesis that the space-time metric, $g_{\mu\nu}$, is a function of the "environment coordinate", τ, in addition to the ordinary space-time coordinates, x^{μ}, where $\mu, \nu = 0, 1, 2, 3$, so that the infinitesimal distance, ds, is given by

$$ds^2 = g_{\mu\nu}(x, \tau) dx^{\mu} dx^{\nu}.$$

Notice that the physical meaning of τ is arbitrary as it can be the temperature (T), the cosmological time(t), a parameter related to the Kaluza-Klein extra dimensions, or anything else. Notice also that τ may not be a continuous valuable but be a discrete number and that there may exist more than one "environment coordinates". I will consider such extensions of the environment-dependent space-time metric later.

In this "more general relativity", let us consider a simple model for general inconstancy given by the generalized Einstein-Hilbert action of

$$S_I = \int d\tau \int d^4x \sqrt{-g}[R(x, \tau) + ...]$$

$$= \int d^4x \sqrt{-g}[2\lambda + (1/16\pi G) R(x) + ...].$$

In this equation we may find the equations of

$$2\lambda = \int d\tau \int d^4x \sqrt{-g}[R(x, \tau) + ...] / \int d^4x \sqrt{-g}$$

and

$$(1/16\pi G) = \int d\tau \int d^4x \sqrt{-g}[R(x, \tau) + ...] / \int d^4x \sqrt{-g} R(x)$$

so that we may obtain the $G-\Lambda$ relation of

$$32\pi G\lambda = \int d^4x\sqrt{-g}R(x)/\int d^4x\sqrt{-g}.$$

Furthermore, by differentiating both hand-sides of the relation with respect to τ, we can obtain the relation between the varying gravitational and cosmological constants:

$$\dot{G}/G = -\dot{\lambda}/\lambda.$$

Since in the homogeneous and isotropic Einstein-Friedmann model of the Universe, the scalar curvature is given by $R = 6[(1-q)H^2 + (k/a^2)]$, where a is the scale parameter in the Robertson-Walker metric and $k = +1, 0, -1$ for the closed, flat, or open universe, the first relation explains the positivity of the cosmological constant, λ, for the present Universe with $-q > 0$ and, probably, $k = 0$. The second relation leads to the prediction of

$$\dot{\lambda}/\lambda = (0.6 \pm 2.0) \times 10^{-12} yr^{-1},$$

from the limit by Thorsett [6]. This prediction is not only consistent with the present experimental observation but also feasible for future experimental tests.

V. Dark Energy, Dark Matter, and Strange Stars - Future Prospects

In conclusion, let us first point out that not only continuous physical constants such as α and G but also discrete physical numbers such as the number of the space-time dimensions n, the number of quark colors N_c, the number of quark-lepton generations N_g, etc. may vary. In fact, an astonishing "dimensional phase transition", which was discussed by myself [36] about three decades ago, may be possible in the history of our Universe. If n is related to N_c as in the "space-color correspondence", which was proposed by myself about four decades ago [37], both of these fundamental physical natural numbers must vary simultaneously.

Before concluding this note, let us introduce two more recent reports on finding an evidence for environment-dependent fundamental physical constants in addition to many recent reports [38]. Very recently, Kanekar et al. have obtained a very strong constraint of $\Delta\alpha/\alpha = (-1.7 \pm 1.4) \times 10^{-6}$ over a look-back time of 6.7Gyrs with HI and OH lines [39]. Even more recently, Webb et al. have found spatial variation of the fine-structure constant with the amplitude of $(1.1 \pm 0.2) \times 10^{-6} GLyr^{-1}$ by using UVES on the VLT [40], which may be taken as a clear evidence for the environment-dependent fundamental physical constant.

Also, as an addendum to my latest note on cosmology in 2014 [41,42], let me propose simple solutions to the most intriguing problems in current cosmology, the problems of the dark energy and the dark matter:

The problem of the dark energy consists of the two questions: 1) "Where does the cosmological constant λ come from?" and 2)"Why is it so large as or so small as $\Omega_\lambda = 1, 2/3$?". A simple answer to the first question is that it comes from the quantum fluctuations of matter as in pregeometry while another to the second is that it is arbitrary since it would not vanish without any miraculous cancellation between the quantum fluctuation and the constant F^{-1} in pregeometry as seen in Section II and since it is time-varying in general inconstancy as seen in Section IV. On the other hand, the problem of the dark matter can be simply solved by either one of the following possibilities: 1) It consists of exotic matter such as strange quark matter in the so-called Bodmer-Terazawa-Witten hypothesis that super-hypernuclear matter consisting of almost equal numbers of up, down, and strange quarks is stable [43, 44]. In fact, many observations of the possible candidates for strange stars consisting of strange quark matter have recently been reported by astronomical experiments [45, 46], inducing many extensive theoretical investigations on strange stars [47]; 2) It consists of "color-balls" [48], the color-singlet complex objects consisting of an arbitrary number of gluons; 3) It consists of **axions** [49]; 4) It consists of WIMPs (Weakly Interacting Massive Particles) such as **superpartners** of the SM (Standard Model) neutrinos, γ, and Z in supersymmetry [50]. Future experiments would tell us which one of these possibilities is a right answer to one of the most mysterious questions in current cosmology.

References and Dictionary

References

[1] Dirac P. A. M., *Nature* **139**, 323(1937); *Proc. R. Soc. London* **A165**, 199(1938); **A333**, 403(1973); **A338**, 439(1974).

[2] Jordan P., *Ann. der Phys.* **6**, *Folge* **1**, 219(1947). For the recent related article, see Schuking E. L., *Physic Today*, October 1999, Vol.**52**, No. 10, p.26(1999).

[3] Hoyl F. and Narlikar J. V., *Nature* **233**, 41(1971).

[4] Canuto V., Adams P. J., Hsieh S.-H., and Tsiang E., *Phys. Rev. D* **16**, 1643(1977).

[5] Terazawa H., *Phys. Lett.* **101B**, 43(1981); in *Proc. 3rd Alexander Friedmann International Seminar on Gravitation and Cosmology*, St.Petersburg, 1995, edited by Gnedin Yu. N., Grib A. A., and Mostepanenko V. M. (Friedmann Laboratory Pub., St.Petersburg, 1995), p.116; *Mod. Phys. Lett.* **A11**, 2971(1996); **A12**, 2927(1997); **A13**, 2801(1998).

[6] Thorsett S. E., *Phys. Rev. Lett.* **77**, 1432(1996). See also Kaspi V. M., Taylor J. H., and Ryba M. F., *Astrophys. J.* **42**, 713(1996) and Guenther D. B., Krauss L. M., and Demarque P., *ibid.***498**, 871(1998). For a recent review, see Uzan J., *Rev. Mod. Phys.* **75**, 403(2003). More recently, Copi et al. have inferred $-4 \times 10^{-13} yr^{-1} < (\dot{G}/G)_{today} < 3 \times 10^{-13} yr^{-1}$ (by assuming a monotonic power law time dependence $G \propto t^{-a}$) from the big-bang nucleosynthesis constraint. See Copi C. J., Davis A. N., and Krauss L. M., *Phys. Rev. Lett.* **92**, 171301(2004). Very recently, Jofré et al. have estimated one of the most restrictive upper limits of $|\dot{G}/G| < 4 \times 10^{-12} yr^{-1}$ by using the surface temperature of the nearest millisecond pulsar, PSR J0437-4715, (inferred from ultraviolet observations) and by considering

only modified Urca reactions. See Jofré P., Reisenegger A., and Fernández R., *Phys. Rev. Lett.* **97**, 131102(2006).

[7] Webb J. K. et al., *Phys. Rev. Lett.* **82**, 884(1999); *ibid.***87**, 091301(2001); Murphy M. T., Webb J. K., and Flambaum V. V., *Mon. Not. R. Astron. Soc.* **345**, 609(2003). Recently, de Souza has claimed that the velocity in which quasars evolve into normal galaxies make us to see the limit of quasars slightly Doppler shifted which explains the "misleading" variation of the fine structure constant. See De Souza M. E., astro-ph/0301085, 6 Jan 2003. However, it seems that his estimate of the shift is overestimated by ignoring the isotropy of quasar expansions. Also recently, Bahcall J. N. et al. have set a robust upper limit on $\dot{\alpha}$, $\dot{\alpha}/\alpha < 10^{-13}yr^{-1}$, by using the strong nebular emission lines of O III. See Bahcall J. N., Steinhardt C. L., and Schlegel D., *Astrophys. J.* **600**, 520(2004). See also Cowie L. L. and Songaila A., *Astrophys. J.* **453**, 596(1995); White R. L., Becker R. H., Fan X., and Strauss M. A., *Astron. J.* **126**, 1(2003); Songaila A., astro-ph/0402347, to be published in *Astron. J.* For a recent review, see Songaila A. and Cowie L. L., *Nature* **398**, 667(1999). Very recently, Tzanavaris et al. have estimated the time variation of $x(=\alpha^2 g_p\mu$ where g_p is the proton g factor and μ is the electron/proton mass ratio, $m_e/m_p)$ from 21-cm and ultraviolet quasar absorption lines over a redshift range $0.24 < z < 2.04$ as $\dot{x}/x = (-1.43 \pm 1.27) \times 10^{-15}yr^{-1}$ for a linear fit. Using the $\Delta\alpha/\alpha$ results and assuming $\Delta g_p/g_p \sim 0$, they have obtained $\Delta\mu/\mu = (2.31 \pm 1.03) \times 10^{-5}$ and $\Delta\mu/\mu = (1.29 \pm 1.01) \times 10^{-5}$. See Tzanavaris P. et al., *Phys. Rev. Lett.* **95**, 041301(2005). More recently, Reinhold et al. have obtained $\Delta\mu/\mu = (2.4 \pm 0.6) \times 10^{-5}$ for a weighted fit and $\Delta\mu/\mu = (2.0 \pm 0.6) \times 10^{-5}$ for an unweighted fit, which indicates, at a 3.5σ confidence level, that μ could have decreased in the past 12 Gyr. See Reinfold E. et al., *Phys. Rev. Lett.* **96**, 151101(2006). Note, however, that the possible variation of μ may not necessarily mean that of the fundamental mass scale in nature but simply do that of the effective electron and/or proton masses due to different environmental parameters such as density, temperature, etc.

[8] Shylakhter A. I., *Nature* **264**, 340(1976); ATOMKI Report No.A/1(1983); Damour T. and Dyson F., *Nucl. Phys.* **B480**, 37(1996); Fujii Y. et al., *ibid.* **B573**, 377(2000), hep-ph/0205206. For the latest reexamination of the Oklo and other constrains, see Fujii Y. and Iwamoto A., *Phys. Rev. Lett.* **91**, 241302(2003). Very recently, Marion et al. have set a stringent upper

bound on $\dot{\alpha}$, $\dot{\alpha}/\alpha = (-0.4 \pm 16) \times 10^{-16} yr^{-1}$, by comparing the hyperfine frequencies of Cs and Rb atoms. See Marion H. et al., *Phys. Rev. Lett.* **90**, 150801(2003). See also Bize S. et al., *ibid.***90**, 150802(2003); Chengalur J. N. and Kanekar N., *ibid.***91**, 241302(2003); Ashenfelter T., Mathews G. J., and Olive K. A., *ibid.***92**, 041102(2004). Even more recently, Fisher et al. have deduced a separate limit of $\dot{\alpha}/\alpha = (-0.9 \pm 2.9) \times 10^{-15} yr^{-1}$ from combining the $1S - 2S$ transition frequency in H atom with the optical transition frequency in Hg^{+} against the hyperfine splitting in Cs and the microwave Rb and Cs clock comparison. See Fisher M., *Phys. Rev. Lett.* **92**, 230802(2004). Very lately, Peik et al. have measured an optical transition frequency at 688THz in $^{171}Yb^{+}$ with a cesium atomic clock at 2 times separated by 28yr and find a value for the fractional variation of the frequency ratio f_{Yb}/f_{Cs} of $(-1.2 \pm 4.4) \times 10^{-15} yr^{-1}$, consistent with zero. Combined with recently published values for the constancy of other transition frequencies this measurement sets an upper limit on the present variability of α at the level of $2.0 \times 10^{-15} yr^{-1} (1\sigma)$ corresponding so far to the most stringent limit from laboratory experiments. See Peik E. et al., *Phys. Rev. Lett.* **93**, 170801(2004). For a recent brief review, see Seife C., *Science* **306**, 793(2004).

[9] Chand H., Srianand R., Petitjean P., and Aracil B., *Astophs.* **417**, 853(2004); Srianand R. *ibid.*, *Phys. Rev. Lett.* **92**, 121302(2004). For a recent comparison of this data with the Oklo data and the quasar data, see Cowie L. L. and Songaila A., *Nature* **428**, 132(2004).

[10] Sakharov A. D., *Dokl. Akad. Nauk SSSR* **177**, 70(1967)[*Sov. Phys. JETP* **12**, 1040(1968)].

[11] Akama K., Chikashige Y., Matsuki T., and Terazawa H., *Prog. Theor. Phys.* **60**, 868(1978); Adler S. L., *Phys. Rev. Lett.* **44**, 1567(1980); A.Zee, *Phys. Rev. D* **23**, 858(1981); Amati D. and Veneziano G., *Phys. Lett.* **105B**, 358(1981).

[12] For a review, see, for example, Terazawa H., in *Proc. 1st Sakharov A. D. Conference on Physics*, Moscow, 1991, edited by Keldysh L. V. and Fainberg V. Ya. (Nova Science, New York, 1992), p.1013.

[13] Landau L., in *Niels Bohr and the Development of Physics*, edited by Pauli W. (Mc Graaw-Hill, New York, 1955), p.52. In 1977, we suggested ab-

normally strong gravity at short distances which breaks some presently reliable basic principles in quantum field theories such as special relativity and microscopic causality at a distance of the order of the Planck length. We also suggested that baryon number generation occurred in the very early Universe, breaking CPT invariance which is based on these basic principles. See Terazawa H., Kasuya M., and Akama K., INS-Report-331 (INS, Univ. of Tokyo, 1979); Terazawa H., in *Proc. Workshop on Unified Theories and Baryon Number in the Universe*, Tsukuba, 1979, edited by Sawada O. and Sugamoto A. (National Lab. for High Energy Physics, Tsukuba, 1979), p.99; Gen.Relativ.Gravit. **12**, 93(1980). For a recent idea similar to the latter suggestion, i.e.,, dynamical breaking of CPT in an expanding Universe and, combined with baryon-number-violating interactions, driving the Universe towards an equilibrium baryon asymmetry that is observationally acceptable, see Davoudiasl H. et al., *Phys. Rev. Lett.* **93**, 201301(2004). For a very recent search for signatures of Lorentz and CPT violations, see Feng B. et al., *Phys. Rev. Lett.* **96**, 221302(2006). Also, recently Calmet et al. have derived fundamental limits on measurements of position arising from quantum mechanics and classical general relativity and shown that any primitive probe or target used in an experiment must be larger than the Planck length. See Calmet X., Graesser M., and Hsu S. D. H., *Phys. Rev. Lett.* **93**, 211101(2004). Very recently, Robinson and Wilczek have calculated the contribution of graviton exchange to the running of gauge couplings at lowest nontrivial order in perturbation theory and found that, when extrapolated formally, it renders all gauge couplings asymptotically free. See Robinson P. and Wilczek F., *Phys. Rev. Lett.* **96**, 231601(2006). Their results may lead to a modified version of the Landau conjecture that the effect of gravity suppresses all gauge couplings at extremely short distances.

[14] Terazawa H., Chikashige Y., and Akama K., *Phys. Rev. D* **15**, 480(1977).

[15] For reviews, see, for example, Terazawa H., in *Proc. 19th International Conf. on High Energy Physics*, Tokyo, 1978, edited by Homma S., Kawaguchi M., and Miyazawa H. (*Phys. Soc. Japan*, Tokyo, 1979), p.617; in *Proc. 22nd International conf. on High energy Physics*, Leipzig, 1984, edited by Meyer A. and Wieczoek E. (Akademie der Wissenschaften der DDR, Zeuten, 1984), vol.I, p.63.

[16] Terazawa H., Chikashige Y., Akama K., and Matsuki T., *Phys. Rev. D* **15**, 1181(1977); Terazawa H., *ibid.***16**, 2373(1977); **22**, 1037(1980); **41**, 3541(E)(1990).

[17] Beringer J. et al.(Particle Data Group), *Phys. Rev. D* **86**, 010001(2012), including the latest world averages of the cosmological data, $\Omega_m = 0.27 \pm 0.03$, $\Omega_\lambda = 0.73 \pm 0.03$, and $t_0 = 13.75 \pm 0.13 Gyr$.

[18] Bekenstein J. D., *Phys. Rev. D* **25**, 1528(1982); Moffat J. W., *Int. J. Mod. Phys.* **D2**, 351(1992); Barrow J. D. and Magueijo J., *Phys. Lett.* **B443**, 104(1998).

[19] Clayton M. A. and Moffat J. W., *Phys. Lett.* **B460**, 263(1999); Albrecht A. and Magueijo J., *Phys. Rev. D* **59**, 043516(1999); Barrow J. D., *ibid.***59**, 043515(19999); Avelino P. P. and Martins C. J. A. P., *Phys. Lett.* **B459**, 468(1999); Davies P. C. W. et al., *Nature* **418**, 603(2002).

[20] Vilenkin A., *Phys. Rev. Lett.* **81**, 5501(1998); Starobinsky A. A., *JETP Lett.* **68**, 757(1998); Pimental L. O. and Diaz-Rivera L. M., *Int. J. Mod. Phys.* **A14**, 1523(1999).

[21] Bjorken J. D., *Ann. Phys.* (N.Y.) **24**, 174(1963); Terazawa H., Chikashige Y., and Akama K., Ref. [14].

[22] Glashow S. L., *Nucl. Phys.* **22**, 579(1961); Salam A. in *Elementary Particle Physics*, edited by Svartholm N. (Almqvist and Wilsell, Stockholm, 1968), p.367; Weinberg S., *Phys. Rev. Lett.* **19**, 1264(1967).

[23] Nambu Y. and Jona-Lasinio G., *Phys. Rev.* **122**, 345(1961); Terazawa H., Chikashige Y., and Akama K., in Ref. [14].

[24] See, for example, Terazawa H., *Phys. Lett.* **133B**, 57(1983).

[25] Kaluza T., Sitzker. *Preuss. Akad. Wiss.* **K1**, 966(1921); Klein O., *Z. Phys.* **37**, 896(1926).

[26] Arkani-Hamed N., Dimopoulos S., and Dvali G., *Phys. Lett.* **B429**, 263(1998); *Phys. Rev. D* **59**, 086004(1999); Antoniadis I. et al., *Phys. Lett.* **B436**, 257(1998); Randall L. and Sundrum R., *Phys. Rev. Lett.* **83**, 3370(1999); **83**, 4690(1999).

[27] Pati J. C. and Salam A., *Phys. Rev. D* **10**, 275(1974); Terazawa H., Chikashige Y., and Akama K., in Ref. [14]; Terazawa H., *Phys. Rev. D* **22**, 184(1980).

[28] For classical reviews, see, for example, Terazawa H., in Ref. [15] and for more recent reviews, see for example, Terazawa H., in *Proc. International Conf. "New Trends in High-Energy Physics"*, Alushta, Crimea, 2003, edited by Bogolyubov P. N., Jenkovszky L. L., and Magas V. V. (Bogolyubov Institute for Theoretical Physics, Kiev, 2003), *Ukrainian J. Phys.* **48**, 1292(2003); in *Proc. XXI-st International Conf. "New Trends in High-Energy Physics"*, Yalta, 2007, edited by Bogolyubov P. N., Jenkovszky L. L., and Magas M. K. (Bogolyubov Institute for Theoretical Physics, Kiev, 2007), p. 271; in *Proc. XXII International Conf. on New Trends in High-Energy Physics*, Alshuta, Crimea, 2011, edited by Bogolyubov P. N. and Jenkovszky L. L. (Bogolyubov Institute for Theoretical Physics, Kiev, 2011), p.352, arXiv:1109.3705v5 [physics.gen-ph] 10 Feb 2012. For our latest papers on unified subquark models of all fundamental particles and forces, see Terazawa H. and Yasuè M., *J. Mod. Phys.* **5**, 2015(2014), arXiv:1401.3562v8[ep-ph] 5 Apr 2014; Nonlinear Phenomena in Complex Systems **19**, 1(2016), arXiv:1505.00172v2[hep-ph] 9 Sep 2015; and Terazawa H., *ibid.* **20**, 327(2017).

[29] For recent more detailed reviews including many other current topics in cosmology, see, for example, Terazawa H., in *Proc. 11th International Conf. and School "Foundations & Advances in Nonlinear Science"*, Minsk, Belarus, 2003, edited by Kuvshinov V. I. and Krylov G. G. (Belarusian State University, Minsk, Belarus, 2004), p.99.

[30] Akama K. and Terazawa H., *Gen. Rel. Grav.* **15**, 201(1983).

[31] For a latest review, see Seife C., *Science* **303**, 1271(2004). Even more lately, an extremely small galaxy has been found with the image taken by the Very Large Telescope. For a brief review, see Seife C., *ibid.* **303**, 1597(2004).

[32] Peacock J. A. et al., *Nature* **410**, 169(2001). For a recent brief review, see, for example, Watson A., *Science* **295**, 2341(2002).

[33] Sievers J. L. et al. (CBI Collaboration), *Astrophys. J.* **591**, 599(2003).

[34] Bennett C. L. et al., *Astrophys. J. Suppl.* **148**, 1(2003); Spergel D. N. et al., *ibid.* **148**, 175(2003). See also Reichhardt T., *Nature* **421**, 777(2003); Carroll S., *ibid.* **422**, 26(2003); Brumfiel G. *ibid.* **422**, 108(2003). For a review, see Schwarzschild B., *Physics Today*, Vol.56, No.**4**, April 2003, p.21. For the most recent and precise data on the cosmological parameters, including $\Omega_m = 0.31 \pm 0.01$ and $\Omega_\lambda = 0.69 \pm 0.01$, see Planck Collaboration: Ade P. A. R. et al., *Planck 2015 Results.XIII Cosmological parameters*, arXiv:1502.01589v2[astro-ph.CO] 6 Feb 2015.

[35] Wheeler J. A., in *Conference on the Role of Gravitation in Physics*, Chapel Hill, North Carolina, 1959 (WADC Technical report 57-216, ASTIA Document No. AD118180; DeWitt B., in *Proc. Third Seminar on Quantum Gravity*, Moscow, 1984, edited by Markov M. A., Berezin V. A., and Frolov V. P. (World Scientific, Singapore, 1985), p.103; Gurzadyan V. G. and Kocharyan A. A., *Zh. Eksp. Teor. Fiz.* **95**, 3(1989)[*Sov. Phys. JETP* **8**, 1(1989); *Mod. Phys. Lett.* **A4**, 50(1989).

[36] Terazawa H., in *Quantum Field Theory and Quantum Statistics*, edited by Batalin I. A. and Vilkovisky G. A. (Adam Hilger, London, 1987), Vol.I, p.637; in **Particles and Nuclei**, edited by Terazawa H. (World Scientific, Singapore, 1986), p.304; in *Modern Problems of the Unified Field Theory*, edited by Golt'sov D. V., Kuzmenkov L. S., and Pronin P. I. (Moscow State University, Moscow, 1991), p.124 (in Russian translation); in *Proc. 5th Seminar on Quantum Gravity*, Moscow, 1990, edited by Markov M. A., Berezin V. A., and Frolov V. P. (World Scientific, Singapore, 1991), p.643. For a recent similar idea of emergence of a four dimensional world from causal quantum gravity, see Ambjørn J., Jurkierwicz J., and Loll R., *Phys. Rev. Lett.* **93**, 131301(2004).

[37] Terazawa H., in Ref. [26]; in *Wandering in the Fields*, edited by Kawarabayashi K. and Ukawa A. (World Scientific, Singapore, 1987), p.388.

[38] For a recent review, see Berengut J. C. and Flambaum V. V., *Nuclear Physics News*, Vol.20, No.3, p.19(2010).

[39] Kanekar N. et al., arXiv:1201.3372v1[astro-ph.CO] 16 Jan 2012, accepted for publication in *Astrophys. J. Letters* (2012).

[40] Webb J. K. et al., *Phys. Rev. Lett.* **107**, 191101(2011); King J. A. et al., arXiv:1202.4758v1[astro-ph.CO] 21 Feb 2012; *Mon. Not. R. Astron. Soc.* **422**, 3370(2012). See also Avelino P. and Sousa L., arXiv:1404.3419v3[astro-ph.CO] 5 May 2015.

[41] Terazawa H., in *Proc. XXI International Seminar on Nonlinear Phenomena in Complex Systems*, Minsk, Belarus, 2014, edited by Babichev L. F., Kuvshinov V. I., and Shaparau V. A. (National Academy of Sciences of Belarus, Minsk, Belarus, 2014), *Nonlinear Dynamics and Applications* **20**(2014), p.241, arXiv:1202.1859v13[physics.gen-ph] 20 Apr 2015, and many references therein.

[42] For a recent review on experimental observations of varying fundamental physical constants, see Stadnik Y. V. and Flambaum V. V., *Phys. Rev. Lett.* **114**, 161301(2015), arXiv:1412.7801v4[hep-ph] 21 Mar 2015; *ibid.***115**, 201301(2015), arXiv:1503.08540v4[astro-ph.CO] 4 Nov 2015. For a recent review on observational probes of cosmic acceleration, see D.H.Weinberg et al., arXiv:1201.2434v2 [astro-ph.CO] 6 Mar 2013.

[43] Terazawa H., INS-Report-336 (INS, University of Tokyo, Tokyo) May, 1979; *J. Phys. Soc. Jpn.* **58**, 3555(1989); **58**, 4388(1989); **59**, 1199(1990). For a classic review on super-hypernuclei, see Terazawa H., in *Proc. 2nd Conf. on Nuclear and Particle Physics*, Cairo, 1999, edited by Comsan N. M. H. and Hanna K. M. (Nuclear Research Center, Atomic Energy authority, Cairo, 2000), p.28. For a more recent review, see Terazawa H., in *Proc. International Conf. on New Trends in High-Energy Physics*, Yalta, Crimea(Ukraine), 2005, edited by Bogolyubov P. N., Fedosenko P. O., Jenkovszky L. L., and Karpenko Yu. A. (Bogolyubov Institute for Theoretical Physics, Kiev, 2005), p.259, arXiv:1304.5655v2 [physics.gen-ph] 15 Apr 2015.

[44] Chin S. A. and Karman A. K., *Phys. Rev. Lett.* **43**, 1292(1979); Witten E., *Phys. Rev. D* **30**, 272(1986); Farhi E. and Jaffe R. L., *Phys. Rev. D* **30**, 2379(1984); **32**, 2452(1985).

[45] For a recent review on exotic nuclei and strange stars, see Terazawa H., *Nonlinear Phenomena in Complex Systems*, **18**, 25(2015), arXiv:1304.5655v2[physics.gen-ph] 15 Apr 2015, and many references therein.

[46] For early proposals of strange stars, see Itoh N., *Prog. Theor. Phys.* **44**, 291(1970); Bodmer A. R., *Phys. Rev. D* **4**, 1601(1971). For a classical review, see Weber F., Schaab Ch., Weigel M. K., and Glendenning N. K., Report No.LBL-37264, UC-413 (LBL, Berkeley, 1995), in *Proc. Ringer Workshop*, Tegernsee, Germany, 1995, presented at Conference: C95-03-06. See also Glendenning N. K., *Compact Stars, Nuclear Physics, Particle Physics, and General Relativity*, 2nd ed.(Springer-Verlag, New York, 2000). For a more recent review, see Weber F., arXiv:astro-ph/0407155v2, 27 Sep 2004, published in *Prog. Par. Nucl. Phys.* **54**, 193(2005). For a latest review, see Weber F. et al., arXiv:1210.1910[astro-ph.SR] 6 Oct 2012, published in *Proc.the IAU Symposium 291*, edited by van Leeuwen J.(International Astronomical Union, 2013), p.61.

[47] For some recent theoretical investigations on strange stars, see Paulucci L. and Horvath J. E., arxiv:1405.1777v1[astro-ph.HE] 7 Mar 2014, *Phys. Lett. B* **733C**, 164(2014); Chu P.-C., Chen L.-W., and Wang X., arXiv:1406.5610v2[nucl-th] 9 Sep 2014, *Phys. Rev. D* **90**, 063013(2014); Biswas S. et al., arXiv:1409.8366v5[nucl-th] 1 Oct 2015, and many references therein.

[48] Terazawa H., *J. Phys. Soc. Jpn.* **69**, 2825(2000); *Nonlin. Phenom. Complex Syst.* **3**, 293(2000), and many references therein. For the latest theoretical investigations on color-balls, see Terazawa H., in Ref.[45] and Terazawa H. and Yasuè M., *Nonlinear Phenomena in Complex Systems* **19**, 1(2016), arXiv:1505.00172v2[hep-ph] 9 Sep 2015, and many references therein.

[49] For some recent theoretical investigations on axions as the dark matter, see Stadnik Y. V. and Flambaum V. V., arXiv:1506.08364v1[hep-ph] 28 Jun 2015; Roberts B. M. et al., arXiv:1511.04098v1[physics.atom-ph] 12 Nov 2015; and many references therein. For a latest review, see Yang Q., arXiv:1509.00673v2[hep-ph] 10 Sep 2015. Very lately, however, Frampton P. H. has disputed whether axions exist. See Frampton P. H., *Searching for Dark Matter Constituents with Many Solar Masses*, arXiv:1510.00400v6[hep-ph] 24 Oct 2015.

[50] Miyazawa H., *Prog. Theor. Phys.* **36**, 1266(1966); Gol'fand Yu. A. and Likhtman L. P., *ZhETF Pis. Red.* **13**, 452(1971)[JETP Lett.**13**, 323(1971)]; Volkov D. V. and Aklov V. P., *ibid.***16**, 621(1972) [*ibid.***16**, 438(1972)]; *Phys. Lett.* **46B**, 109(1973); Wess J. and Zumino B., *Nucl. Phys.* **B70**,

39(1974). Very lately, Frampton P. H. has queried arguments for WIMPs which arise from electroweak supersymmetry and argued that dark matter constituents must uniquely be primordial black holes if they constitute all dark matter. See, for the details, Frampton P. H., in Ref.[49]. For earlier proposals of primordial black holes as dark matter constituents, see Terazawa H., in Ref. [45].

Dictionary

Einstein-Hilbert action of gravity: The action in which the Lagrangian density contains the scalar curvature term and the ordinary matter term so that the Euler equations of motion derived from the minimum action principle may lead to the familiar Einstein equations of motion for gravity.

Kaluza-Klein extra dimension: The fifth dimension (in addition to the ordinary four dimensions of space-time) which was hypothesized by Kaluza in 1921 and by Klein in 1926 in their unified gauge theory of gravity and electromagnetism[25].

large number hypothesis: The hypothesis which states that the Eddington large numbers, $N_1 = \alpha/Gm_em_p = O(10^{39})$, $N_2 = m_e/\alpha H = O(10^{40})$, and $N_3 = 4\pi\rho/3m_pH^3 = O(10^{80})$ (where α, G, and H are the fine-structure, gravitational, and Hubble constants; m_e and m_p are the electron and proton masses; and ρ is the baryon density of the Universe, respectively), are not independent but related with each other as $N_2 = c_2N_1$ and $N_3 = c_3(N_1)^2$ where c_2 and c_3 are constants of order close to unity, which was proposed by Dirac in 1937[1].

Maxwell action of electromagnetism: The action in which the Lagrangian density contains the term of electromagnetic field strength squared so that the Euler equations of motion derived from the minimum action principle may lead to the familiar Maxwell equations of motion for electromagnetism.

momentum cut-off of the Pauli-Villars type: The cut-off of momentum in the Feynman loop-integration by introducing the hypothetical particles with the same quantum numbers as the ordinary particles in the loop, but with much larger masses(called "cut-off mass or energy") and the negative metric.

new large number relations: The less large number relation stating that the total number of galaxies in the Universe(N_G) and the average total number of stars in a galaxy(N_S), both of which have the same order of magnitudes of 10^{11}, are not independent but related; and the even less large large number relations stating that the three ratios of R_U/D_G. D_G/R_G, and R_G/D_S(where R_U, D_G,

R_G, and D_S are the "radius of the Universe", the average distance between two neighboring galaxies, the average radius of a galaxy, and the average distance of two neighboring stars), all of which have the same order of magnitudes of 10^3, are not independent but related with each other. Both of these relations were hypothesized by the author in 1995[5].

pregeometric phase: The phase of space-time in which the vacuum expectation value of the space-time metric either diverges or vanishes so that there is no physical space-time where particles can move as in the geometric phase in which the vacuum expectation value of the space-time metric neither vanishes nor diverges, and is finite and well-defined.

superpartners: The hypothetical particles with the same quantum numbers as ordinary particles except for the spins different by $h/4\pi$ from the spins which ordinary particles have. They make a super-doublet with the ordinary particles. The superpartners of the photon, W, Z, and H bosons are called "photino", "wino", "zino", and "Higgsino".

Postscript

In this book, I have attempted to introduce to you the most advanced way of understanding the nature from subquarks to the universe in the unified composite model of all fundamental particles and forces. I hope that even high-school students are able to read the contents in Chapter I. *Quark Matter and Strange Stars* without so much difficulty by consulting with the Dictionary for Chapter I while college or undergraduate students may be able to understand the contents in Chapter II. *Composites of Subquarks as Quark Matter*, by consulting with the dictionary in Chapter II. However, I am afraid that they may have some difficulty in reading a part of the contents in Chapter III. *Dark Energy, Dark Matter, and Strange Stars*, even after consulting the dictionary in the chapter, as it contains some sophisticated mathematical equations which only graduate students in mathematics or physics major may be able to understand.

After reading this book, some of you may have the impression that the current physic is so much advanced for one to be able to understand almost everything in the universe that there may not be so many subjects left for further investigations. At the end of this book, however, I would like to remind you of the following episodes on the lives of two eminent physicists living in the twentieth century:

Dr. Robert Oppenheimer made such great scientific contributions to the fields ranging from atoms in the material to black holes in the universe, but, regrettably, created nuclear (or Atomic) bombs in the United States. Also, Dr. Andrei Sakharov proposed the pregeometric theory of gravitation, the three necessary conditions for explaining the baryon asymmetry in the universe, and the "muon catalyzed fusion" but, regrettably, created thermo-nuclear (or Hydrogen) bombs in the Soviet Union.

I suppose that we, physicists, will keep going on to research in this twenty-first century, in order to find definite answers to many fundamental questions such as "What is matter made of ?" and "How does it interact?", "How was the universe created?" and "Whether will it end?", and so on, in contemplating what the purpose of physics is!

About the Author

The author, *Hidezumi Terazawa*, was born in Hyogo, Japan, on July 25, 1942. He graduated from the University of Tokyo (BS in 1965, MS in 1967, and DS in 1970), and became an instructor-research associate at the Laboratory of Nuclear Studies, Cornell University in 1969, an assistant professor of The Rockefeller University in 1971, an associate professor of Institute of Nuclear Study, University of Tokyo in 1975, Founding Director-in-general of CAOS (Center of Asia and Oceania for Science) in 1989, and Founding and Life-long Advisor of MABT (Midlands Academy of Business and Technology) in 2003.

Index

M

N

P

Q

S

T

V

W

Photoluminescence: Advances in Research and Applications

Editor: Ellis Marsden

Series: Physics Research and Technology

Book Description: In this collection, chalcogenide glasses doped with rare earth elements are proposed as particularly attractive materials for applications in integrated photonics. The opening chapter is dedicated to reviewing the studies on optical properties of $(GeS_2)_{100-x}(Ga_2S_3)_x$ ($x=20$, 25 and 33 mol%) glasses, doped with Er_2S_3 in a wide range from 1.8 to 2.7 mol%, by absorption and photoluminescence (PL) spectroscopy.

Hardcover ISBN: 978-1-53613-537-4
Retail Price: $160

Ionizing Radiation: Advances in Research and Applications

Editors: Tamar Reeve

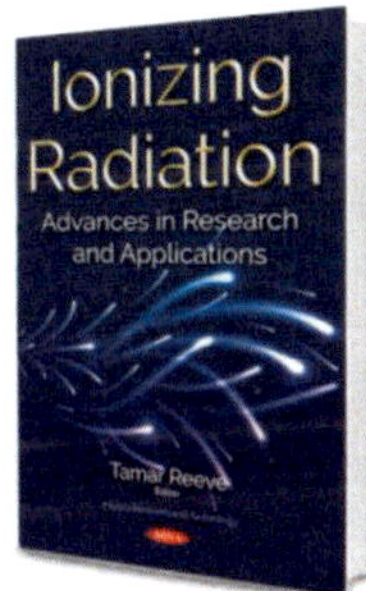

Series: Physics Research and Technology

Book Description: In this compilation, the authors examine the importance of ionizing radiations for thermoluminescence dosimetry that is the current area of research for medical and industrial purposes. Ionizing radiations are harmful to the human body, so, there is a need to measure small doses in the environment as well as very high doses at the time of accident like radiation leakage and for the treatment of cancer.

Hardcover ISBN: 978-1-53613-539-8
Retail Price: $195

Models of Plasma Kinetics and Problems with Their Interpretation in the Current Paradigm

Authors: Vladimir V. Lyahov and Vladimir M. Neshchadim
(Leading Research Workers of Institute of Ionosphere, SLLP

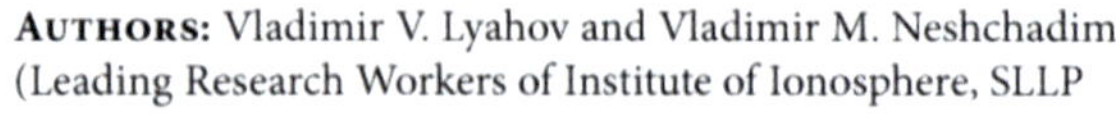

Series: Physics Research and Technology

Book Description: Proposed by A.A. Vlasov in 1938, the kinetic equation with a self-consistent electromagnetic field led to a fundamentally new perspective in plasma physics. This equation represents the most profound approach to the description of plasma because it operates directly with plasma particles using the distribution function.

Hardcover ISBN: 978-1-53612-853-6
Retail Price: $195

Perturbation Theory: Advances in Research and Applications

Editor: Zossima Pirogov

Series: Physics Research and Technology

Book Description: *Perturbation Theory: Advances in Research and Applications* begins with a deliberation on the development of a formalism of the Exchange perturbation theory (EPT) that accounts for the general identity principle of electrons that belong to different atomic centres. The possible applications of the theory concerning scattering and collision problems are discussed, and the authors apply the TDEPT to the description of the positron scattering on a Lithium atom as an example.

Hardcover ISBN: 978-1-53612-989-2
Retail Price: $195